Grooming, Gossip, and the
Evolution of Language

Grooming, Gossip, and the Evolution of Language

ROBIN DUNBAR

HARVARD UNIVERSITY PRESS
Cambridge, Massachusetts

Library of Congress Cataloging in Publication Data

Dunbar, R. I. M. (Robin Ian MacDonald), 1947–
Grooming, gossip, and the evolution of language / Robin Dunbar.
p. cm.
Includes bibliographical references.
ISBN 0-674-36334-5 (cloth)
ISBN 0-674-36336-1 (pbk.)
1. Human evolution. 2. Social evolution. 3. Language and
languages—Origin. 4. Gossip—History. 5. Human behavior.
6. Group identity. 7. Interpersonal relations. I. Title.
GN281.4.D85   1996
573'.2—dc20   96-15934

Contents

Acknowledgements

This book inevitably owes a great deal to a great many people. I am grateful to all those with whom I have discussed the ideas it contains, and especially so to those who have contributed to the research on which it is based. Particular thanks go to: Leslie Aiello, Rob Barton, Dick Byrne, Richard Bentall, Hiroko Kudo, Peter Kinderman, Chris Knight, Sam Lowen, Dan Nettle, Sanjida O'Connell, Boguslaw Pawlowski and Peter Wheeler. Neil Duncan, Amanda Clark, Nicola Hurst, Catherine Lowe, David Free and Anna Marriott helped with various research projects, and Nicola Koyama tracked down the references for the bibliography. As always, I am indebted to my editor Julian Loose for his enthusiasm and patience.

RD

CHAPTER I

Talking Heads

To be groomed by a monkey is to experience primordial emotions: the initial frisson of uncertainty in an untested relationship, the gradual surrender to another's avid fingers flickering expertly across bare skin, the light pinching and picking and nibbling of flesh as hands of discovery move in surprise from one freckle to another newly discovered mole. The momentarily disconcerting pain of pinched skin gives way imperceptibly to a soothing sense of pleasure, creeping warmly outwards from the centre of attention. You begin to relax into the sheer intensity of the business, ceding deliciously to the ebb and flow of the neural signals that spin their fleeting way from periphery to brain, pitter-pattering their light drumming on the mind's consciousness somewhere in the deep cores of being.

The experience is both physical sensation and social intercourse. A light touch, a gentle caress, can convey all the meanings in the world: one moment it can be a word of consolation, an apology, a request to be groomed, an invitation to play; on another, an assertion of privilege, a demand that you move elsewhere; on yet another, a calming influence, a declaration that intentions are friendly. Knowing which meaning to infer is the very basis of social being, depending as it does on a close reading of another's mind. In that brief moment of mutual understanding in a fast-moving, frenzied world, all social life is distilled in a single gesture.

To recognize what this simple gesture signals in the social world of monkeys and apes, you need to know the intimate details of those involved: who is friends with whom, who dominates and who is subordinate, who owes a favour in return for one granted the week before, who has remembered a past slight. The very complexity of the social whirl creates those ambiguities we are so familiar with from our own lives.

Take, for instance, Jojo, who has just given birth to her first offspring. She cradles it in her arms, at once puzzled by this strange, wet thing and unsure what she should do. Already alert, the baby struggles to turn its head, as though surprised by the unfamiliar sights and sounds that surround it. They are not alone for long. Jojo's mother, Persephone, comes across. She peers down at the baby, sniffs it tentatively, and reaches out a hand to touch its rump. Persephone gives a quiet grunt and begins to groom Jojo, leafing through her fur, busying herself with the rituals of social interaction. But she cannot take her mind off the baby and keeps pausing to reach down and groom its head briefly, making smacking noises with her tongue and lips as she does so.

Jojo relaxes to the rhythm of her mother's grooming, and her eyes half close. But she jolts awake again when her baby gives a whimper. Two young juveniles are poking at the infant, fascinated by its wriggling as they pull tentatively on a leg. Jojo pulls the baby away and turns her back on them, disrupting Persephone's grooming. Persephone stares meaningfully at the two juveniles, her head lowered and her eyebrows raised in threat. The juveniles scamper off to annoy someone else.

Jojo and Persephone happen to be baboons, members of a troop whose life centres on a rocky outcrop in the wooded grasslands of eastern Africa. But they could be almost anywhere in Africa. Indeed, they could be members of any one of about a hundred and fifty species of monkeys and apes that live in the forests and woodlands of Asia, Africa and South America. Moreover, there is something eerily familiar in the process of their actions and responses – as if they might be humans too, members of any one of some 5000 cultural groups scattered across the globe from Alaska to Tasmania, Benin to Brazil. Here, in the minutiae of everyday life, is a point of convergence between ourselves and our nearest relatives, the monkeys and apes. Here is behaviour with which we instantly empathize, the innuendoes and subtleties of everyday social experiences.

Yet there is one difference: our world is infused through and through with language, while theirs goes about its business in wordless pageant.

A human baby produces its first real words at about eighteen months of age. By the age of two, it has become quite vocal and has a vocabulary of some fifty words. Over the next year it learns new words daily, and by the age of three it can use about 1000 words. It is now stringing words together in short sentences of two or three words, calling your attention to objects, requesting this and that. Its command of grammar is already nearly as competent as an adult's, though it will still make amusing yet plainly logical mistakes, saying *'eated'* instead of *'ate'*, *'mouses'* instead of *'mice'*. Then the floodgates open. By the age of six, the average child has learned to use and understand around 13,000 words; by eighteen, it will have a working vocabulary of about 60,000 words. That means it has been learning an *average* of ten new words a day since its first birthday, the equivalent of a new word every 90 minutes of its waking life.

This is an extraordinary achievement. It is no wonder that the machinery which makes this possible is so expensive to maintain. Although your brain accounts for no more than 2 per cent of your body weight, it consumes 20 per cent of all the energy you eat. In other words, pound for pound, the brain burns up ten times as much energy to keep itself going as the rest of the body does. The situation is even more extreme in young children, where the brain is actively growing as opposed to just ticking over. During the last stages of pregnancy, the foetus's brain is growing very rapidly and consumes 70 per cent of the total energy the mother pumps into her baby via the umbilical cord – and she, of course, has to provide all that. Even after birth, the brain still accounts for 60 per cent of the infant's total energy consumption during the first year of life. Lactation is an exhausting business.

It will come as no surprise to discover that we humans have the largest brains relative to body size of any species that has ever existed. Our brain is nine times larger than you would expect for a mammal of our body size. It is thirty times larger than the brain of a dinosaur of the same body size. Only the porpoises and dolphins come close to us in this respect; yet even though dolphins are renowned for their intelligence and sociability, they still do not compete with humans on the verbal scale. Complex though their natural language of whistles and clicks may be, it does not seem to be in the same league as human language.

*

Because it seems to be unique, language appears all the more miraculous. Other species bark and scream, grunt and wail, but none speak. Perhaps inevitably, this has encouraged us to view the human species as special, reinforcing our habits of self-worship. Yet, when we look at our nearest relatives, the monkeys and apes, we find much that is familiar – the same intensity of social life, the petty squabbles, the joys and frustrations, the same whining children irritating exasperated parents as in our own private lives. However, neither monkeys nor apes have language in any sense that we would recognize from our everyday experience of human conversation.

How did it come about that we, the descendants of just such dumb apes, have this extraordinary power when they do not? The puzzle seems all the greater because we feel so at home with the social lives of monkeys and apes. What makes it seem familiar to us is the time they spend in close physical contact, busily attending to each other's needs in endless grooming sessions. They think nothing of spending hours leafing through each other's fur, combing, picking, parting the hairs with the single-mindedness of a human mother attending to her child's tangled mop.

The answer to this apparent puzzle lies, I suggest, in the way we actually use our capacity for language. If being human is all about talking, it's the tittle-tattle of life that makes the world go round, not the pearls of wisdom that fall from the lips of the Aristotles and the Einsteins. We are social beings, and our world – no less than that of the monkeys and apes – is cocooned in the interests and minutiae of everyday social life. They fascinate us beyond measure.

Let me give you a few statistics to reinforce the point. Next time you are in a café or a bar, just listen for a moment to your neighbours. You will discover, as we have in our research, that around two-thirds of their conversation is taken up with matters of social import. Who is doing what with whom, and whether it's a good or a bad thing; who is in and who is out, and why; how to deal with a difficult social situation involving a lover, child or colleague. You may happen on a particularly intense exchange about a technical problem at work or a book just read. But listen on, and I'll wager that, within five minutes at the most, the conversation has drifted away again, back to the natural rhythms of social life.

Even were you to listen to the conversations in university common rooms or the restaurants of multinational companies, there at the very hub of our intellectual and business life the situation would not be all that different. To be sure, you will occasionally come across an intense discussion of some abstruse scientific technicality or business deal. But that will only be when a visitor is being entertained or when individuals meet for the specific purpose of thrashing out some key problem of mutual concern. For the rest of the everyday conversations, it's unlikely that more than about a quarter would be concerned with matters of such intellectual weight as the cultural, political, philosophical or scientific issues of the day.

Here are two more statistics, this time gleaned from the world of the printed page. Of all the books published each year, it is fiction that tops the list in terms of volume of sales. Take a glance around your local bookshop: university campus bookshops aside, two-thirds of the shelf space will contain fiction. Even then, it is not the rip-roaring adventure yarns that attract us, but the unfolding intimacies of the main characters. It is the way they handle their experiences that fascinates us, their reactions to the vagaries of life – those 'there but for the grace of God go I' situations. And out of all this fiction, it is not the writings of the acclaimed masters that top the publishers' sales-lists, but romantic fiction.

Everything else – from art history to photography and sport, from the sciences and handicrafts to car mechanics for the home enthusiast – is put together under the all-encompassing label of 'non-fiction'. Only biographies can lay any claim to a significant share of the market in their own right. Every year, a torrent of such books appears, retelling the life experiences of the rich, the famous and the also-rans. Every TV newscaster, every politician, every actress, every minor sportsman from darts to football, has published his or her story. Long-dead novelists, generals and politicians all command their fair share of attention.

And why do we buy such books? It's not to learn about the sport in question, or how to read the news on TV, but to learn about the private lives of our heroes or those who have become as familiar to us as our own families. We want the intimate details, the gossip, their innermost thoughts and feelings, not detailed technical analysis of method acting or parliamentary procedure. We want to know how events affected them, how they reacted to the highs and lows of life, what they thought about their friends and relations, the indignities they suffered, the triumphs they took part in.

Take another look at your daily paper. How many column-inches are devoted to substantive news about politics and economics? Here is the score for two of yesterday's papers, the upmarket London *Times* and the mass-market UK tabloid *The Sun*. No less than 78 per cent of the 1063 column-inches of text in the downmarket *Sun* was concerned with 'human interest' stories, stories whose sole purpose seemed to be to enable the reader to become a voyeur of the intimate lives of other individuals. That leaves just 22 per cent for news and commentary on the political and economic events of the day, for the sports results, for news of upcoming cultural events, and for everything else. Even the august *Times* only devoted 57 per cent of the 1993 column-inches of text in its main news and review sections to political and technical news; 43 per cent was devoted to human interest stories (interviews, news stories of a more salacious kind, and so on). The number of actual column-inches devoted to 'gossip' was virtually identical in the two papers: 833 and 850 respectively.

It's clear that most of us would rather hear about the doings of the great and the not-so-good than about the intricacies of economic processes or the march of science. The trial of O. J. Simpson aroused more interest and achieved higher viewing figures than the deliberations of US congressional committees, despite the fact that the conclusions of those committees will have an impact on our future lives far beyond any conceivable impact that OJ's guilt or innocence might have.

Here, then, is a curious fact. Our much-vaunted capacity for language seems to be mainly used for exchanging information on social matters; we seem to be obsessed with gossiping about one another. Even the design of our minds seems to reinforce this. Of course, great things are possible with language: Shakespeare and T. S. Eliot are not figments of our imagination, neither are the unsung writers of instruction manuals; we really can use language for profit and pleasure. And language remains our greatest treasure, for without it we are confined to a world that, while not one of social isolation, is surely one that is a great deal less rich. Language makes us members of a community, providing us with the opportunity to share knowledge and experiences in a way no other species can. So how is it that we have this extraordinary ability, yet most of the time seem to do so little with it?

A century of intensive research in linguistics, psychology and speech science has taught us a great deal about language: how it is produced, what grammar does, how children learn it. Yet at the same time, we know almost nothing of why it is that we alone, of all the tens of thousands of living species, possess this extraordinary ability. We do not know for sure when it evolved or what the first languages ever spoken sounded like. However, during the last ten years we have learned more about the background to human evolution and the behaviour of our nearest relatives, the monkeys and apes, than we had in the previous thousand years put together; and this new evolutionary perspective, firmly rooted in modern Darwinian biology, has focused our attention on questions about language that have hitherto been overlooked. In the process, aspects of our own past that had been buried beneath the murky waters of history for hundreds of millennia have finally come to light.

The approach I adopt is thus very different from the perspectives of those who study language conventionally. For the past century, the study of language has focused principally on three main areas: linguistics (with its pervasive interest in the structure of grammars); socio-linguistics (with its interest in the way sex and social class influence the words we use and how we pronounce them); and the neurobiology of language (the brain structures that allow us to speak and understand). Although there has been some interest in the archaeology and the history of language (and the processes of dialect formation), these have been considered both peripheral and trivially speculative by the mainstream interests.

Even less attention has been devoted to the function of language and the reasons why we have it and other species do not. Indeed, such questions have often been deliberately eschewed. Instead, language has often been viewed as an 'epiphenomenon', something that appeared as a by-product of other biological processes (notably, our super-large brains) and for which no other kind of explanation is necessary.

This curious state of affairs owes its origins largely to the claim (widely held in the social sciences) that human behaviour in general, and language in particular, are social phenomena and thus lie beyond the pale of biological explanation. Neurobiology might provide us with insights into the machinery of language production and comprehension, but beyond that, biology sheds little light on the nature of language. By and large, biologists have respected this demarcation line. But the recent developments in evolutionary biology have had far-reaching implications for our understanding of human behaviour as well as that of other animals, and language has inevitably come under this new and more powerful microscope.

This book is about those new discoveries, and about the origins of our capacity for language. I shall examine not only what we do with language but also the more fundamental questions of why we have it, whence it came and how long ago it appeared. The story is a magical mystery tour that will take us bouncing from one unexpected corner of our biology to another, from history to hormones, from the very public behaviour of monkeys and apes to the moments of greatest human intimacy. It will take us back through the chapters of human history to the time before we were human, when we were but apes of a not especially unusual kind. What did the earliest languages sound like? Who spoke them? And why, from these early hesitant steps, did languages evolve, changing and diversifying so much that now we have around 5000 mutually incomprehensible tongues (and that's not counting the ones that became extinct in the millennia before anyone could write them down)?

CHAPTER 2

Into the Social Whirl

What characterizes the social lives of humans is the intense interest we show in each other's doings. We spend literally hours in each other's company, stroking, touching, talking, murmuring, being attentive to every detail of who is doing what with whom. You might think that this marks us out as a cut above the rest of life, but you would be wrong. If we have learned anything from the last thirty years of intensive research on monkeys and apes, it is that we humans are anything but unique. Monkeys and apes are just as social as we are, just as intensely interested in the social whirl around them.

So to set the scene on the human story, we need to go back in time to our primate heritage. What is it about primates that makes them so different from other animals, that in turn gives us our unique character? The answer is that primates live in a very much more sophisticated kind of social world than other animals do.

The Monkey on my Back

Monkeys and apes are highly social species. Their lives revolve around a small group of individuals with whom they live, work and have their being. Without its friends and relations, a monkey would no more be able to survive than a human being could. The social life of primates is intense and all-consuming. They spend a great deal of the day engaged in social grooming with their special friends. Like Jojo and Persephone, whose story opened Chapter 1, these are often matrilineal relatives, related through their mother's line in an unbroken chain of personalized mother-daughter relationships that run back through the mists of time to some ancestral primate pre-Eve.

The biologist Richard Dawkins has reminded us just how short this chain of ancestry really is. Imagine yourself, he says, standing on the Indian Ocean shore just where Kenya abuts on to the southern border of Somalia. Face south and reach out to hold your mother's left hand in your right hand. Facing you is a chimpanzee of the same age and sex, holding its mother's hand in its left hand. Your mother is holding her mother's hand in her right hand, and the chimpanzee's mother is holding *her* mother's hand in her left hand. The double chain of generations snakes its parallel way across the African plains westwards towards the distant peak of Mount Kenya, a faint brown smudge emerging above the clouds on the horizon. By the time the chain reaches Mount Kenya itself, a distance of no more than 300 miles, the mother-daughter lines have converged and met in a single mother-Eve. She lived somewhere on the East African savannahs some time between 5 and 7 million years ago.

The number of generations between you and this ancestral Eve is surprisingly small. Even allowing a modest yard and a half for the span of outstretched arms and twenty years as the average generation length, there are no more than 350,000 individuals separating you on the Kenyan coast from her on the slopes of Mount Kenya. That's barely a third of the people who work for the National Health Service in the UK, no more than the total population of a modest English county town or, to put it into really dramatic perspective, about half the number of babies born in England and Wales each year. Even allowing just ten years per generation (probably a better estimate of the typical age at which females give birth for the first time among chimpanzees and our earliest ancestors), the line of life would stretch no further than the western shore of Lake Victoria, some 600 miles from the coast, perhaps 700,000 individuals in all.

It's a sobering thought that so few generations separate us from the common ancestor we share with the chimpanzees. Here, indeed, is not just our cousin but our sister-species. It is no wonder that some biologists have started to refer to us humans as the third chimpanzee (in addition, that is, to the common chimpanzee and its closely related sister species the bonobo or pygmy chimpanzee).

But let's pursue Dawkins' graphic metaphor a little further back in time. How much further will we need to go to reach the common ancestor of the Old World monkeys and apes?

At most 85 miles further on, a week's easy walk beyond Mount Kenya, we come to the common ancestor of the gorilla and the chimp-human lineage. That's something like 100,000 generations if females give birth for the first time at about ten years of age, as most great apes do. On the same scale, some 700 miles further on we come to the common ancestor of the human-chimp-gorilla family and the orang-utan, the endearing red ape of the Asian forests. We are now just on the Uganda-Zaïre border, a mere stone's throw to the north of the Virunga Volcanoes where Dian Fossey lived and died watching her beloved mountain gorillas.

Generation lengths get shorter as we go further back among the smaller-bodied species, perhaps 5-6 years on average once we are past the common ancestor of the living species of great ape – but at the same time, the length of an outstretched arm is now no more than 18 inches, a yard between adjacent noses. It's just another 400 miles to the common ancestor of the great apes and the gibbons, the lesser apes now found only in south-eastern Asia. From there to the common ancestor of all the monkeys and apes of Africa and Asia will be another 1100 miles. By now, we are somewhere in the middle of Congo-Brazzaville, with still another 500 miles to go to the Atlantic Ocean. We have not even traversed the African continent at its narrower part. We have travelled back in time through 30 million years, and there have been just four million females in the unbroken chain that links you and that long-dead pre-pre-Eve that gambolled through the tree-tops in an ancient African forest. That's less than half the population of London or Paris, barely a quarter of the population of modern-day Rio de Janeiro.

We are almost exactly half-way back in time towards the Age of Dinosaurs. We are still a long way from the origins of the first, primitive pre-primates in the dying days of these great reptiles' empire. Most people are surprised to find that, so far from being a new and advanced product of evolution, the primates are in fact one of the oldest lineages of mammals, a close relative of the insectivores, the bats and the flying foxes. Their and our early ancestors dodged the same heavy-footed lizards in the twilight years of the long reign of the dinosaurs.

The ancestral primates were small, squirrel-like animals with long pointed noses that scuttled among the bushes and trees of the dense tropical forests that existed during the closing millennia of the Age of the Dinosaurs. In the new freedom brought by the post-dinosaur years, they diversified into myriad new species in hundreds of different niches, mostly in the northern hemisphere in what is now Europe and North America. These species were all prosimians, whose living descendants include the lemurs of Madagascar and the galagos and lorises of Africa and Asia. For 30 million years they dominated the woodlands and forests of the northern hemisphere.

Then the earth's climate cooled rapidly over a period of around two million years. Water surface temperatures in the tropical Pacific dropped from a sultry 23 °C to around 17 °C.[1] The tropical zones shifted southward towards their present equatorial position. As these climatic changes developed, one of these prosimian groups began to evolve in an entirely new direction. Brain size increased, faces became more rounded. It was the beginning of a major break with the past that ultimately gave rise to the so-called anthropoid primates (the monkeys and apes as we know them

1. We can determine temperatures in the remote past thanks to a curious quirk of physics. There happen to be two isotopes (or forms) of oxygen, one being fractionally heavier than the other. A molecule that happens to have been formed with atoms of the heavier isotope, oxygen-18, will not evaporate from the oceans as easily as its counterparts made from the lighter oxygen-16 atoms; but once evaporated, it condenses more rapidly to form snow. Hence, by measuring the ratio of the two forms of oxygen in ice or snow, we can tell whether one time period was colder or warmer than another: colder periods will have a higher proportion of the lighter oxygen isotope. Ice cores drilled from the Greenland or Antarctic ice caps are examined inch by inch for changes in the ratio of the two oxygen isotopes. With their ratio in samples of today's snow as a standard, the series can be anchored to current global temperatures, and past temperatures can be read off. Another way of doing it is to examine the composition of the calcium carbonate in the shells of extinct oceanic plankton. The plankton take up oxygen from the seas in which they live to make their skeletons. In cooler times, there will be a higher proportion of the heavier oxygen isotope in sea water because more of the lighter ones will have evaporated.

today). By this time, primates were confined pretty much to their present distribution in the equatorial regions of Africa, Asia and South America. Soon afterwards, contact between Africa and South America was lost. The South American monkey populations went their own way, evolving into species that are still reminiscent of the ancestral anthropoid primates of 35 million years ago.

In Africa and Asia, meanwhile, evolution proceeded apace. Around 30 million years ago, this branch of the primates split into two major families, the Old World monkeys (now represented by colobus and langur monkeys, baboons and macaques) and the apes. It was to be the apes that dominated the forests of the Old World for the next 15–20 million years, however. The monkeys remained relatively insignificant.

Some time around 10 million years ago, the forests of the Old World began to retreat as the climate started to dry out and temperatures once more plummeted. Surface temperatures on the earth's oceans dropped by another 10°C. Within a few million years the apes had succumbed, and the monkeys, better adapted to a terrestrial way of life and able to out-compete the apes by eating poorer-quality diets, came into their own.

Part of the problem seems to have been that, unlike the monkeys, apes lack the ability to neutralize the tannins in unripe fruit. Tannins are poisons that plants produce to make their parts inedible. Mature leaves often contain high concentrations of tannins, apparently to prevent herbivores from stripping the leaves off a tree and thereby killing it. But sometimes animals can be useful to plants. Rooted to the spot, plants have a problem about dispersing their seed. It doesn't pay to have all your children growing under your feet, because they simply compete with you and each other for sunlight and the nutrients in the soil. It does pay to disperse your seeds widely, where the young plants will compete with (and, you hope, out-produce) other plants' offspring. The problem that plants face is how to achieve this dispersal.

Their saviours, if you like, were animals. Mammals such as monkeys can travel several miles in a day. By harnessing their energies, a plant can disperse its seeds over a very wide area indeed. Species like figs, plums and apples provide their seeds with a luxurious coating of energy-rich flesh to entice animals to eat the seed. The seed passes slowly through the animal's gut (a passage that can take two or three days, by which time the animal may be several miles away), and gets dumped to germinate far from its parent.

Plants that adopt this strategy have a problem, however. The seed has to develop to its final form and size to be able to germinate under its own steam. All the nutrients required for the early stages of germination have to be provided by the mother plant. To prevent fruit-eating animals from destroying the immature seed before it can fend for itself, plants that produce these kinds of seed protect their fruits with tannins and other compounds. These poisons gradually break down as the fruit ripens, so that by the time the seed at the centre of the fruit is ready to cope with its great journey through life, the flesh that surrounds it has shed its chemical defences. It is the tannins that give unripe fruit that bitter, mouth-drying taste.

We share with the great apes an inability to digest unripe fruit. Lacking the enzymes that break down the tannins, we suffer a stomach-ache or, in the worst cases, diarrhoea if we eat too much of it.

However, some time during their evolutionary history the Old World monkeys acquired the enzymes and other mechanisms that allow them to beat the chemical defences of plants. Being able to eat unripe fruit may well have given monkeys like the baboons and macaques a distinct edge over the ape lineages as things began to get tough in the forests seven million years ago. With significant quantities of fruit being eaten before it ripened, there was much less available for the apes. Gradually but inexorably, the apes went into decline and the monkeys took over as the dominant primates of the forest. Those few species of apes that survived were forced into increasingly marginal habitats like the forest floor and the forest edge, where monkeys seldom ventured. Today, the remnants of that once-successful ape lineage cling to survival by their fingernails, confined to small pockets here and there in Africa and Asia, their populations dwindling by the decade.

Meanwhile, whole new groups of monkey species appear for

the first time in the fossil record and soon begin to dominate the scene. The macaques, now confined to Asia but once widespread throughout Europe and north Africa, appear some 10 million years ago. The baboons appear a few million years later. The guenons are more recent still: the earliest recognizable guenons are barely two million years old – younger than the first truly recognizable members of our own genus, *Homo*, who inhabited the lake and river margins of eastern Africa some two and a half million years ago.

Most of us naturally assume that the monkeys of Africa and Asia represent the ancestral condition through which we all passed in this long story. Traditional conceptions of the evolution of the primates envisaged a natural progression by which the common monkey of the Old World transformed itself first into apes and then into modern humans. Most people are surprised to discover that this is incorrect. The new science of molecular genetics, combined with a better understanding of anatomy and a newly discovered wealth of fossil material, has revealed that the familiar monkeys of Africa and Asia, like the baboons, guenons and macaques, are in fact upstart newcomers by comparison with the apes, the lineage to which we humans belong.

The disaster that hit the luckless apes 10 million years ago was not, of course, complete and final. One lineage of apes did survive the depredations of the climatic changes: the lineage that eventually led to us. Some time around 7 million years ago, one population seems to have begun to make increasing use of the more open savannahs that bordered the forest to which the apes were (and still are) so well adapted. They were probably forced to do so by their failure to compete successfully with the other ape species that were vying with increasing desperation for the ancestral forest-home. But as so often happens in evolutionary history, the challenge of exploiting a marginal habitat forced the pace of evolutionary change. Mortality would have been desperately high, but those that survived did so because they were able to exploit the new conditions. In that crucial moment, a mere blink of an eye on the geological time-scale, our history hung in the balance between extinction and survival. It must have been touch and go.

Despite their very different and at times turbulent histories, all these lineages have faced the same agonizing problem: how to survive the depredations of a seemingly endless catalogue of predators ranging from sabre-tooth cats to lions and leopards, from hyenas and hunting dogs to monkey-eating eagles and occasionally even other primates. Finding food is, of course, a perennial problem for any animal, but given enough time it can almost always scratch a living from the natural world. The problem is that, as it does so, it inevitably exposes itself to the risk of being taken unawares by a predator. In order to gain the advantage of surprise over their prey, most predators rely on stealth before throwing themselves into the final chase. So every minute spent travelling from one feeding site to another, every minute spent with its attention focused on removing a fruit or a leaf from the branch in front of it, exposes an animal to the risk of being caught by a predator lying in wait for an inattentive prey.

The intensity of predation depends on your size. For species as large as chimpanzees and gorillas, the risks of predation are much reduced (though by no means altogether absent). But for smaller species it can be an ever-present threat. It has been estimated that about a quarter of all vervet monkey deaths are due to predation (mostly by leopards), while as many as 20 per cent of all red colobus monkeys at the Gombe National Park in Tanzania fall prey to hunting by Jane Goodall's famous chimpanzees.

Predation is a significant evolutionary problem, because animals that find themselves on the inside of a predator no longer have the opportunity to breed and reproduce. Since animals that fail to reproduce do not contribute any of their genes to future generations, there is intense pressure to find ways of circumventing this unhappy fate. Evolution is the outcome of successful solutions to the problem. Indeed, the very fact that you and I are alive today is a consequence of the remarkable circumstance that every one of our ancestors – back to, and far beyond, that ancestral pre-pre-Eve – successfully solved the problem of survival, at least for long enough to reproduce.

They did so by exploiting two main facts about predators. One is that a predator cannot easily handle a prey animal significantly larger than itself. Only a handful of specialized, group-hunting predators can do this successfully. Hence by increasing your body size, you considerably reduce your vulnerability to predators. Terrestrial species are more vulnerable to predators than arboreal species because they have less opportunity to escape into the dense foliage or fine outer branches of trees where pursuit becomes too risky for predators. As a result, all the terrestrial species are larger than their arboreal counterparts.

The other way of reducing the risk of predation is to live in large groups. Groups reduce the risk in a number of ways. One is simply by providing more eyes to detect stalking predators. Most predators have to get within a certain distance of their prey undetected in order to have any chance of catching it. That's why your cat inches its way, belly-down in the grass, towards the birds pecking at the breadcrumbs you so kindly threw on to your lawn. Using every blade of grass and every molehill as cover, the cat moves only when it is sure the birds are not looking, freezing in its tracks the moment it senses the birds might be suspicious.

Each predator has its own attack distance, depending on its speed and its style of attack. For a cheetah, capable of hitting 70 miles an hour within seconds from a standing start, the attack distance is 70 yards; for the slower and more bulky lion it is 30 yards, while for the lighter leopard it is just 10 yards, often less. If prey can detect a stalking predator before it can get within its attack distance, they will always be able to outrun the predator. Most predators are well enough aware of that, if only by virtue of past experience, and rarely bother to chase a prey that has already seen it. That's one reason why you will occasionally see a lion walking through a herd of wildebeest with the herd simply parting like the biblical Red Sea around the advancing predator. The wildebeest know that so long as they stay outside the lion's attack distance they are relatively safe and need do no more than keep a wary eye on it.

Larger groups are also an advantage as a deterrent. Most predators will be less enthusiastic about attacking a prey animal if they know that several others will come to the victim's aid. Although it is virtually unknown for species like wildebeest to go to the aid of a fellow herd-member brought down by a lion or hunting dogs, group defence is more common among primates. Baboons have been known to drive leopards up a tree, and have occasionally even killed them. Red colobus monkeys are significantly less likely to be attacked by chimpanzees if an adult male of their group is nearby; even chimps seem unwilling to risk a mass counter-attack launched by an animal that's barely a quarter of their own weight. Think of it in human terms. Being handbagged by one granny would not put off the average mugger, but even the most determined thug will balk at being handbagged by twenty grannies simultaneously.

Last but not least, a group creates confusion in a predator. Predators succeed by locking on to a target animal and running it down. When the prey runs into a group, there are animals running every which way and the predator becomes momentarily confused. That moment of lost attention is often just enough for the prey to make good its escape.

So primates live in groups as a mutual defence against predation. Indeed, sociality is at the very core of primate existence; it is their principal evolutionary strategy, the thing that marks them out as different from all other species. It is a very special kind of sociality, for it is based on intense bonds between group members, with kinship often providing a platform for these relationships. Primate groups have a continuity through time, a history built on kinship (usually mother-daughter relationships, but very occasionally father-son ones too).

A Friend in Need

Living in groups creates its own tensions, as any member of a close-knit community knows only too well. There is that once-in-a-while thoughtless transgression on your personal space, the time when someone treads on your tail in the mêlée around a particularly attractive food source. Or worse still, the time when some other scoundrel unashamedly steals the very food from your mouth just as you settle down to feed. These are the everyday trials of social life, the hassles of crowding so familiar to city commuters and the inhabitants of inner-city high-rise tenements, the frustrations of chronically overcrowded housing and large families. They are the centrifugal forces that drive us all apart in search of the peace and serenity of being alone.

Social animals hang in perpetual balance between two forces: the centripetal forces, driven by fear of predation, which have produced the feelings of sociableness that make us seek out company; and the centrifugal forces, generated by overcrowding, that send us scurrying for the sanity of a solitary life. When predators become common (and for humans, those predators can just as easily be neighbouring human groups rampaging through your territory on raids), we hanker for the close proximity of friends and tolerate all kinds of overcrowding. When predators are rare, the stresses of crowding overwhelm us and we disperse. Group size is the product of this balancing act.

Primates have evolved a distinctive response to this problem. Large, tightly bonded groups are their solution to the risks of predation; but in order to be able to increase group size, it was first necessary to develop a mechanism that enabled them to keep fellow group members just far enough away so that they don't become too much of a nuisance, while at the same time not driving them away altogether. The fine balance is brought about by coalitions between small numbers of animals. A mother and her daughters, or two sisters perhaps, form an alliance in which they provide mutual support against everyone else. It's an 'if you'll help me, I'll help you' arrangement. It seems to be unique to the higher primates. Although male lions form coalitions to take over prides of females, these are often temporary affairs, formed on the spur of the moment with a specific end in view. The coalitions of monkeys and apes are long-term commitments, often formed months ahead of their being needed. They are a promise of future action in circumstances as yet unimagined.

Of the thousands of hours I have spent watching monkeys in Africa, perhaps the most unashamedly enjoyable were those spent studying an obscure but unusually attractive species of baboon found only on the high mountains of the Ethiopian plateau. This is the gelada, once also known by the evocative name of the bleeding heart baboon because of the hourglass-shaped patch of bare red skin on its chest. The males are truly magnificent, sporting lion-like capes of long russet and black hair that hang from their shoulders and float on the breeze as they run along the cliff-tops. Gelada live in harems that typically consist of four or five females, their dependent offspring and a breeding male (the harem male). While sons usually leave to join an all-male bachelor group soon after puberty, daughters mature into the group to join their mothers, older sisters, aunts and female cousins in a coalition of great intensity and loyalty. In effect, these alliances are formed at birth, the product of being born to a particular mother.

These deeply rooted alliances have important implications for the harem males. They are perpetually at risk of being displaced by younger males anxious to get a foot on the ladder of immortality by finding females with whom they can breed. Because there are four or five females in each harem but only one breeding male, many of the males in the population are excluded from breeding. These form the core of the all-male groups, biding their time in anticipation of a suitable opportunity. Sooner or later, desperation forces their hand and they make a bid to take over a harem of females. Needless to say, the incumbent harem males are less than enthusiastic about the prospect of losing their harems – for once the harem is lost, they have no further opportunity to mate and breed.

In an effort to forestall the inevitable, harem males devote a great deal of their time to trying to scare off the opposition on the sidelines with impressive displays. At the same time, they try to ensure that their females do not stray too far from them lest they find an opportunity for illicit dallyings with these males. Whenever a female wanders too far away, or inadvertently finds herself too close to another male (even another harem male), her male will warn her with raised eyebrows and panted threats. Occasionally, these may escalate into a charge, culminating in a vigorous display of threats over the cowering female.

The harem male's attempts to ride herd on his females when they stray too far from him often backfire at this point. The luckless victim's grooming partners invariably come to her aid. Standing shoulder to shoulder, they outface the male with outraged threats and furious barks of their own. The male will usually back off and walk huffily away, endeavouring to maintain an air of ruffled dignity. Occasionally, however, the male will persist, feeling perhaps unusually sensitive about his honour or security. This only leads to more of the group's females racing in to support their embattled sisters. The male invariably ends up being chased round the mountainside by his irate females in an impressive display of sisterly solidarity.

These alliances are established and maintained by grooming, the most social activity in which monkeys and apes engage. In some species, as much as a fifth of the entire day may be spent grooming, or being groomed by, other group members. A mother will spend hours devotedly grooming her offspring, carefully leafing through its fur in search of dead skin, matted hair, bits of leaf and burrs that have become entangled in its hair as the animal brushes its way through the vegetation during the day's foraging. She will also groom her friends and relations, in what seems to be selfless devotion to their hygienic interests. Keeping the fur clean and the skin healthy is obviously an important factor in the life of any animal.

But there is more to grooming than just hygiene, at least in the monkeys and apes. For them, it is an expression of friendship and loyalty. Robert Seyfarth and Dorothy Cheney of the University of Pennsylvania spent the better part of a decade during the late 1970s and early 1980s studying the vervet monkeys that live in Kenya's Amboseli National Park. In one of their studies, they recorded the screams uttered by individual vervets that were being attacked by another member of the group. Then, when the screaming animal was physically out of sight, they played these calls back over loudspeakers hidden in bushes, at the same time recording on video the responses of the target animals sitting immediately in front of the loudspeaker.

When they played the calls for help back to most of the animals in the group, they elicited little response other than a cursory glance in the direction of the hidden speaker. But when they played the calls back to an animal that the caller had groomed with during the previous two hours, that animal immediately looked up and stared into the bushes. It was as though it was try-ing to make up its mind whether to go and investigate more close-ly, in order to fulfil its obligation to a grooming partner. Did the situation warrant help, or was it a minor spat that would quickly blow over?

The vervets clearly differentiated between the animals they groomed regularly and those they didn't. A grooming partner is something special, someone who deserves particular attention, who should be supported in moments of need, on whose behalf the taking of risks is warranted.

Gelada operate in the same way. Even on the small scale of the harem unit, the females are very selective about those they groom and those whom they support in altercations. The frequencies with which females are supported (both when they are involved in squabbles within their own unit and when they are attacked by members of another harem for accidentally transgressing into their space) correlate with the frequency with which they are groomed. They are clearly well aware of whom they owe loyalty to, and they don't have to have groomed with them half an hour beforehand to know it.

Not all primate societies exhibit these characteristics, it must be said. Prosimians like the lemurs of Madagascar and the galagos of Africa rarely exhibit coalitionary behaviour, even when they live in groups. And while not unknown, it is by no means common among South American monkeys and some lineages of Old World monkeys like the colobus. Those species that show coalitionary behaviour at its most highly developed tend to be those like the baboons, macaques and vervet monkeys, and the chimpanzee, that live in relatively large groups.

Enter Machiavelli

Coalitionary behaviour seems to be possible only because the ani-mals understand how other individuals tick and how they rate as allies against possible opponents in the group. These are the kinds of knowledge that cannot always be acquired firsthand. Fights are not so common in a primate group that you would be able to see every potential ally fighting against everyone else in turn. Instead, the monkeys seem to weigh up the possibility that if Peter can beat Jim and Jim can defeat Edward, then it's very likely that Peter would defeat Edward should they ever come to blows.

This kind of inference about social relations, combined with an acute sense of others' reliability as allies, seems to be the foundation on which primate alliances are built. At the cognitive level, they are really quite sophisticated kinds of social inference to make. But given that you *can* make them, a new possibility presents itself. Knowledge can be put to bad use, as in propaganda, as well as good. Monkeys and apes use their social skills to exploit each other.

A classic case of this was observed by Andrew Whiten and Dick Byrne during their studies of chacma baboons in southern Africa. A young adult female named Mel was digging a succulent tuber out of the ground. It was a particularly tough job, and one far beyond the strength of all but an adult animal. But the prize of a nutritious tuber in the impoverished habitat these animals inhabit is well worth the effort. Meanwhile, a young juvenile named Paul had been quietly watching Mel at work. Just at the crucial moment when Mel managed to wrench the tuber clear of the ground, Paul let out an ear-splitting scream, of the kind commonly uttered by juveniles who are being attacked by someone much bigger and stronger than themselves. Paul's mother, who had been busy feeding out of sight on the far side of some bushes, immediately came racing through. She took in the situation at a glance, added two and two and made five: Mel had obviously threatened her darling little boy. She fell on the unsuspecting Mel with the kind of enthusiasm uniquely characteristic of mothers whose children are being molested. Needless to say, the startled Mel dropped her tuber and fled, with the outraged mum in hot pursuit. Paul nonchalantly picked up the abandoned tuber and settled in to enjoy lunch.

Observations of this kind are not uncommon. The Swiss zoologist Hans Kummer described a case in which a young female hamadryas baboon spent twenty minutes edging her way inch by inch across a distance of just two yards to get to a rock behind which lay the group's young male follower. Once there, she began to groom with him, sitting upright with the top of her head in full view of her harem male a few yards away. Hamadryas have a similar social structure to the gelada, with small harems of two or three females monopolized by a single breeding male. One key difference between the two species, however, is that hamadryas males are completely intolerant of their females grooming (or even being near) other males. It seemed as though the female had engineered herself into a position where her male would be left with the impression that she was engaged in some completely innocent activity.

In his book *Chimpanzee Politics*, the Dutch zoologist Frans de Waal describes a classic example of the subtle balancing act that the more advanced primates can sometimes engage in. In the captive group of chimpanzees housed at Arnhem Zoo in the Netherlands, the young male Luit had just overthrown the old male Yeroen. Yeroen had been dominant male for some years and during that time he had enjoyed more or less exclusive access to the females when they were ready to mate. But his fall to second rank meant that he lost his privileges to Luit. Worse was to follow a few months later, when the young male Nikkie reached the point where he was able to defeat Yeroen too. Yeroen sank to third rank and lost all privileges. Then came the stroke of genius: rather than bemoaning his bad luck and getting depressed, the wily old male formed an alliance with the young Nikkie. Being younger than Luit, Nikkie was no match for the new dominant male on his own. But with Yeroen's support, he was able to defeat Luit. The result was a new ranking, with Nikkie at the top and Yeroen second, and Luit squeezed back into third place.

Then came the *coup de grâce*. Yeroen proceeded to make use of his position to mate with the females. Nikkie, of course, took instant umbrage and set about chastising the presumptuous Yeroen. Yeroen patiently bided his time. On the next occasion when Nikkie and Luit got into a squabble, Yeroen simply sat on the sidelines and refused to go to Nikkie's help. So Nikkie lost the battle, and would have lost the war had he not quickly settled his difference with Yeroen. So long as he tolerated the old male mating with at least some of the females, things worked out just fine.

But every time Nikkie forgot himself in his jealousy, Yeroen would remind him by withdrawing his support against Luit.

These manipulations are only possible because monkeys and apes are able to calculate the effect their actions are likely to have. This does not of course mean that Yeroen was working out the odds on his pocket calculator; indeed, it is not even clear in this case how much of Yeroen's success came from deliberate scheming and how much from luck. We can stand outside the events and, with hindsight, read a cunning plan into the story as it unfolds. But when we ourselves are embroiled in the action, we rarely think about it in quite such clinical terms. Like Yeroen, we act more by instinct, sensing an opportunity to be gained on the hoof. Still, there was a consistency to Yeroen's behaviour that implies some kind of forethought, even if only at the superficial level of recognizing the opportunities that circumstances pushed his way. The fact that similar behaviour has been observed in wild chimpanzees by the Japanese primatologist Toshisada Nishida lends credence to the view that chimps at least are capable of seeing the implications of their actions and factoring these into their plans for the future. Nishida gave the rather evocative name 'alliance fickleness' to Yeroen's kind of manipulation of Nikkie.

There is considerable evidence to suggest that monkeys and apes are sensitive to the risks they run in these kinds of situation and adjust their behaviour accordingly. Saroj Datta has shown that juvenile female rhesus monkeys are significantly less likely to rush into supporting an ally against a higher-ranking opponent when the opponent's mother is nearby than when she is out of sight. It is as if they know that the opponent's mother is unlikely to sit by while they flatten her precious offspring – and, worse still, that high-ranking mothers tend to have large numbers of relatives who are easily drawn into a squabble in defence of their collective status. There is no point in helping a friend if by so doing you merely exacerbate the situation and cause both of you to end up being flattened.

I have seen gelada behave in a similar way. One day, a young female was attacked by her harem male for straying too far away from the rest of the group. He stood over her, threatening and grinding his teeth in high dudgeon. The female's mother was feeding about five yards away at the time. She looked up the moment the commotion started, but made no effort to intervene. Eventually, his point made, the male turned and stalked off to start feeding a few yards away. As the victim walked disconsolately back towards the rest of the group, her mother called to her with a soft grunt. The victim at once turned and walked over to her mother, who then began to groom her. I had the distinct impression that the mother did not want to become involved in a squabble with the male (perhaps because she sensed that doing so would simply escalate the fight), but at the same time she realized that failing to do so had weakened her relationship with her daughter. The grunt and the grooming seemed to say, 'I'm sorry!'

Frans de Waal has described similar behaviour in chimpanzees and macaques, and has termed it 'reconciliation'. You might think of it in terms of an apology designed to restore the status quo in an alliance that has been damaged by the thoughtless behaviour of one of its members. In most cases, reconciliation involves grooming, touching or other physical actions. Chimpanzees kiss each other on the lips; macaques groom or hold another's rump in a half-mount; male baboons will reach through to touch another male's penis. However, Joan Silk, Dorothy Cheney and Robert Seyfarth have recently reported vocal forms of reconciliation, similar to those just described in the gelada, in chacma baboons inhabiting the Okavango swamps of Botswana. They found that dominant females will give conciliatory grunts when approaching lower-ranking females with whom they want to interact. More importantly, they are more likely to grunt before approaching a female they have threatened earlier than one they haven't. It's as though they want to say, 'Don't worry, my intentions are strictly friendly.'

Reconciliations may also occur between males at times when alliances form an important part of their social strategies. When a male gelada acquires a breeding group of females by defeating an incumbent harem male, his position is far from secure. He was only able to wrest the harem away from the former incumbent because the females' loyalty to that male was weak. The two males will have fought a veritable Battle of the Titans, and probably have sustained serious injuries from each other's two-inch-long canine teeth in the process. But despite the intensity and bravado of the fight, the decision on who wins and who loses lies in fact in the hands of the females; it is they who ultimately decide whether or not to desert their current male in favour of the rival, and they may decide in his favour even though he has been losing the long-running battle with the rival. The problem for the rival if he wins is that, since the females are clearly willing to desert one male in favour of another, they may be just as willing to desert the second, should he prove no more to their taste than the first. There are always plenty of onlookers at a take-over battle who would be willing to give it a go at the slightest hint that the loyalty of the females might be in doubt. And it does sometimes happen.

Caught in this awful bind, victorious males work fast to establish an alliance with their defeated predecessor. The two males may have spent a whole day, sometimes two, locked in intermittent and often bloody combat, but once the decision has been made and the defeated male has accepted the verdict, the winner sets about building a new relationship with him. This involves making a number of tentative overtures. The new male approaches the defeated male in a non-aggressive, almost submissive way. The defeated male is, of course, suspicious at first. He has just received the beating of his life at the hands of this thug. He is sore, exhausted, and nervous of another unprovoked attack. But he wants to stay on in the unit because the current batch of infants represents his last throw of the reproductive dice, and he would like to see them through at least to the point where they can survive on their own.

So the two males have interests in common: the new male would like to have the old male's support against further take-over attempts (at least over the initial period while his position is still uncertain), and the old male would like to stay on in order to protect his offspring. After one or two false starts, the deal is struck surprisingly quickly; there is a simple ritual of reconciliation in which the old male reaches through to touch the new male's penis as the latter presents his rear. Then the two males groom each other with the kind of enthusiasm reserved for the aftermath of patched-up quarrels. And after that, the two are as inseparable as twins when it comes to defending the females against outsiders.

It is the subtleties and complexities of these interactions that give the societies of monkeys and apes their special quality. We can watch the soap opera of daily life unfolding before us, and empathize with the stratagems and counter-stratagems as they evolve. It all looks so familiar, so reminiscent of everyday life in our own societies. This is our primate heritage, our common evolutionary experience. It has important implications for the way our minds are designed, and so in turn for the design of our brains.

A Darwinian Detour

When Charles Darwin published his landmark book *On the Origin of Species* in 1859, he set in motion a revolution that has radically altered our understanding of the living world. So it seems odd, nearly one hundred and fifty years later, to be reminding you that the natural world is a Darwinian world. Yet the lessons of Darwin's revolution in the biological sciences remain widely misunderstood, not just by the layman but also by many scientists outside organismic biology.[2] With even scientists con-

2. The biological sciences can be divided into three rather broad levels. Organismic (or whole organism) biologists study the emergent properties of an animal's behaviour: this includes such topics as as ecology, animal behaviour, population biology and evolutionary processes. Infra-organismic biologists study the processes that make an animal tick: among these can be numbered physiology, cell biology, anatomy and embryology. Finally, molecular biologists study the chemical processes that produce organisms: this newest and in some ways most successful of the sub-disciplines of biology focusses on the way DNA and the other components of the genetic system build cells in the great miracle of life. Although all three layers of biological science are united into a single structure by Darwin's theory of evolution by natural selection, they do not all need to worry about its details to the same extent. While organismic biologists would find it very difficult (if not impossible) to do their research in the absence of Darwin's theory, infra-organism and molecular biologists are much less dependent on it (and indeed, a handful of them even manage to adopt an overtly anti-Darwinian position without compromising their science). A theory of evolution is just not needed to study how a cell works, even though it is all but imposssible to understand the behaviour of the organism of which the cell is a part without the benefit of evolutionary theory. Laymen are sometimes confused by this, assuming that the fact that a cell biologist can do without Darwin must mean that Darwin is wrong. Cell biologists undoubtedly do

fused, it's small wonder that (according to the latest surveys) 48 per cent of the population of the USA still believes that the biblical story of Genesis is literally true.

This all sits very incongruously with the fact that Darwinian evolutionary theory[3] is widely recognized to be the second most successful theory in the history of science (after modern quantum physics). Not only has it been extraordinarily effective in explaining why the biological world is as it is, but it has also proved remarkable in its ability to continue generating new questions to stimulate and guide empirical research. Yet, a century and a half after Darwin first proposed his theory, most people's views of the biological world are still heavily coloured by opinions that were current during the eighteenth century, long before Darwin was even born.

Darwinian evolutionary theory is so fundamental to our understanding of the events I describe in this book that I am obliged to pause in my story to ensure that we all start with the same clear understanding of what the Darwinian account entails. The popular literature is so full of misconceptions and, in some cases, plain fictions that misunderstandings can easily arise. (Readers who feel thoroughly at home with the modern Darwinian perspective may wish to skip the rest of this chapter and go straight to the next.)

One trivial example is the fact that most people are surprised to discover that Darwin did not invent the theory of evolution. In fact, biologists had come to accept the idea of evolution long before Darwin published his seminal work. The second half of the eighteenth century had witnessed a number of significant challenges to the overpowering influence of the biblical world view. One was the realization that the diversity of life on the planet

better biology with the benefit of Darwinian theory as a working framework, but they can get by quite adequately without it, at least for the time being. Whether, as our knowledge of cell biology grows, they will always be able to get by without it remains to be seen.

3. The modern theory is Darwinian but it is not, strictly speaking, Darwin's. It contains many elements that Darwin would not have recognized and could not have known about. Over the last 150 years, biologists have built on Darwin's original insights to produce what, by any standards, is one of the most remarkable and comprehensive theories in science.

could be more easily explained as a consequence of evolution than by the conventional biblical creation story. The first concerted attempt to provide a general theory of evolution was made as early as 1809 by the renowned French biologist Jean Baptiste de Monet, Chevalier de Lamarck – more commonly known to succeeding generations of biologists as plain Lamarck. Darwin's contribution to this debate was not to prove the theory of evolution, but to provide a mechanism – natural selection – that could explain why evolution took place.

Lamarck's views were premised on the Aristotelian *scala natura*, the 'scale of nature', sometimes known as the Great Chain of Being. This curious hangover from the ancient Greeks supposed that all life formed a natural hierarchy which began with things like insects and worms – the original Greek version actually started with earth and water on the lowest rungs of all – proceeded through more advanced forms like fishes, reptiles and birds, to end with mammals, humans and, at the pinnacle of all, the gods. Adopted more or less wholesale by the early Christian Church – with the angels and, on the final rung, God himself substituting for the ancient deities – the Great Chain of Being coloured the way everyone in post-medieval Europe thought about the biological world. Lamarck and his contemporaries built these ideas into their theories of evolution, supposing that each species begins life on the lowest rungs of the ladder and, over long periods of time, gradually progresses up through the hierarchy of life in response to a natural unfolding of some inner force.

Darwin turned all this on its head by insisting that there was no natural progression up an evolutionary ladder. In fact, no species – not even humans – could be considered as better or worse than any other. There is only one biological standard by which a species can be judged, namely its success at reproducing itself. We are all, bacterium and human alike, equally 'good' because we are each sufficiently well adapted to our particular circumstances to thrive and reproduce. The fate of all species is eventually either to become extinct or to be transformed into new species. But in either case, it is natural selection – reflected in the individual organism's ability to survive and, more importantly, reproduce –

that drives these changes, not some internal biological principle or 'life force' as Lamarck assumed.

Darwin's theory has two important lessons for us when we come to think about behaviour. One is that evolutionary change is driven by animals' need to adapt to changing circumstances. The geological sciences have revealed that the earth's climate has been in a constant state of flux, oscillating between the overheated and the frigid with almost monotonous regularity. In the 65 million years since the dinosaurs became extinct, for example, average global temperatures have dropped by an astonishing 18°C. Even the Antarctic was once clothed in dense forest. Associated with these climatic changes have been dramatic changes in vegetation and fauna.

Most of these shifts in climate have been triggered by the changing shape and distribution of the continental masses as they have slithered around on the surface of the planet's soft inner core. Other factors have also played a role, including long-term changes in the distance between the earth and the sun and in the tilt of the earth's axis. Interspersed throughout the 450 million-year history of life on earth there have been five (possibly six) episodes of mass extinction when virtually all existing life forms were wiped out.[4] The handful of survivors provided the seed corn for a whole new series of evolutionary developments that took life on earth off in an entirely new and often quite arbitrary direction.

The way in which climatic changes drive evolution is nicely illustrated by the fact that deep-sea creatures like sharks (whose origins predate the dinosaurs) have remained virtually unchanged for hundreds of millions of years, whereas species like antelopes and humans (both of whom are of very recent origin) have changed dramatically in appearance over a very short period of time. Unlike land habitats, the deep-sea environment tends to be much less

---

4. There is growing (but still controversial) evidence that these mass extinctions were associated with the impact of comets or large asteroids. The dust and water vapour thrown up into the atmosphere by these collisions is presumed to have blocked out the sunlight and created what amounts to a 'nuclear winter'. The last of these mass extinctions occurred 65 million years ago and saw the demise of the dinosaurs.

affected by global temperature changes; consequently, deep-sea creatures face much the same conditions now as their ancestors did 200 million years ago. In contrast, the environments faced by land animals have changed dramatically as a result of sometimes catastrophic changes in the world's climate and vegetation.

Darwin's theory can account for changes of this kind because, unlike Lamarck's species-based theory, it assumes that the individual is the basic unit of evolution. It is the individual that reproduces or doesn't reproduce, and the individual that passes on its particular traits.[5] Where earlier biologists viewed species as ideal types (clones, you might say), Darwin and his colleagues began to see the species as simply a collection of sometimes quite variable individuals who shared a number of key traits. That variation was the potential that allowed species to evolve, though evolution would only occur *if* natural selection made change advantageous.

The other key lesson of the Darwinian approach is that in real life nothing comes for free. Evolutionary change is not automatic; it always occurs against a gradient of stability imposed by the organism's natural biological coherence. Change along one dimension of an organism's biology (let's say growing taller) always incurs costs. One reason is that the change may throw other aspects of the system out of kilter: a taller individual tends to be more gangly, and so less fast at escaping predators. A second problem is that any change (such as growing taller or developing a bigger brain) costs energy: bigger individuals need more food to fuel their bigger bodies. For evolution to occur, the benefits to be gained from changing a character have to exceed the costs. When there is no advantage to change, the costs of making changes act as a stabilizing force, selecting for constancy. To understand evolutionary change, we have to understand both the costs and the benefits of any particular course of action.

The one exception to this arises in those cases where the charac-

5. Strictly speaking, as Richard Dawkins has reminded us in his book *The Selfish Gene*, it is the gene that is the fundamental unit of evolution: evolution occurs because certain genes are passed on to the next generation more successfully than other genes. However, speaking of individuals rather than genes is a convenient shorthand.

ters concerned are protected from the immediate impact of the forces of selection. This happens when different versions of a gene produce the same effect. When there is no selection pressure for or against a change, genetic traits can drift as a result of chance events that affect which individuals happen to reproduce. Consequently, a population that splits into two halves which remain reproductively isolated for a long enough period will accumulate minor genetic differences. The theory of neutral selection, originally promulgated by the Japanese geneticist Motoo Kimura in the 1970s, has proved a valuable tool for evolutionary biologists because it allows us to determine the length of time since any two species had a common ancestor. It is simply a matter of determining the number of mutations by which their DNA differs and multiplying this by the average rate at which spontaneous mutations occur. This is how we know that humans and chimps shared a common ancestor who lived some time around 5 to 7 million years ago.

One final point needs to be emphasized. Many people find Darwinism disturbing because they confuse it either with Social Darwinism, with its associations with the eugenics movement of the early 1900s, or with genetic determinism. The first of these confusions is frankly bizarre, because despite its name Social Darwinism had very little to do with Darwinism; it was largely the brainchild of the social philosopher Herbert Spencer, aided and abetted by the distinctly anti-Darwinian founder of genetics Francis Galton (ironically a cousin of Darwin's). Irrespective of whether Darwin was himself a Social Darwinist, the movement's basic philosophy – maintaining the purity of the species – was distinctly Lamarckian in conception. Indeed, by the 1920s it became apparent that Darwinism had pulled the intellectual rug well and truly from under it. The Social Darwinists (and their curious afterimage, the Nazis in the 1930s) were motivated by fears that the excessive fertility of the socially unfit underclasses was diluting the viability of the human species. In fact, the underclasses were behaving in a respectably Darwinian fashion: reproducing as fast as they could to ensure that their genes made it into the next generation, despite the high mortality rates their children suffered thanks to their grinding poverty. If anything, they were dutifully increasing the range of diversity on which natural selection has to work, thereby reducing the likelihood of our species' extinction in the long term. Heaven forbid that we should all have ended up as clones of the upper classes!

The bugaboo of genetic determinism is for many a modern-day version of Social Darwinism. But once again, the problem is largely one of misinformation – sometimes exacerbated by a refusal to listen. Evolutionary biology has no preconceptions about the genetic determinism of behaviour, even though at some point genes must be involved. Many features of an animal's behaviour can be explained in terms of strategies that maximise genetic fitness, even though the behavioural rules they use are learned or culturally inherited. Learning is just another example of a Darwinian process: it is differential survival of traits (behavioural rules in this case) as a result of selection. When animals make decisions about how to behave, they do so on the basis of past experience, and in the light of the costs and benefits of particular courses of action. Their decisions may well be guided by genetically instilled intuitions about how fitness can be maximized, but they do not act blindly in response to inner drives beyond their control; advanced organisms can act or hold back, according to circumstances. Every day of their lives animals make decisions about whether the risks entailed in behaving in a particular way are worth what they will gain by it.

So much for our Darwinian detour. Now back to monkeys.

CHAPTER 3

The Importance of Being Earnest

What makes groups of primates different from groups of other species is their 'busy-ness'. Every waking moment has something of significance going on. Here is a grooming, there a squabble that is sorted out by an ally, elsewhere a subtle deception – the whole welded together by a constant watchfulness, taking in who-is-doing-what-with-whom. At the root of it all, however, are the long sessions of grooming so peculiarly characteristic of primate societies. Here, in some imperfectly understood way, lies the key to the processes that give primate societies their cohesion and sense of belonging.

A Sense of Touch

Grooming takes up a great deal of a monkey's time. In most of the more social species, the grooming of other individuals accounts for about 10 per cent of an animal's day. But in some species, it can take up as much as 20 per cent of the animal's time. That is an enormous commitment, given that finding food is a time-consuming activity at the best.

We know that grooming is intimately related to an animal's willingness to act as an ally of another individual. At least among the Old World monkeys and apes, the time devoted to grooming during the day correlates roughly with the size of the group. This makes a certain amount of sense. If grooming is the cement that holds alliances together, then the more time you devote to grooming your ally, the more effective that alliance will be. And since alliances will be proportionately more important to you the larger the group gets, it makes sense to invest even more time in grooming your allies. But why it should be so effective in this respect remains unclear.

Prosimian primates (lemurs and galagos) also spend a great deal of time grooming, and Rob Barton of Durham University has shown quite convincingly that, when lemurs groom, it is mainly for reasons of hygiene. Grooming by other individuals tends to concentrate on those parts of the body (the scalp and back) that an animal cannot reach for itself. Social grooming is in these circumstances a helpful tit-for-tat arrangement – quite literally a case of 'you scratch my back and I'll scratch yours'.

At one level, grooming is simply a pleasurable experience. Studies of captive monkeys have shown that grooming makes them more relaxed, reducing their heart rate as well as other external signs of stress. They sometimes become so relaxed that they fall asleep. In fact, we now know that grooming stimulates the production of the body's natural opiates, the endorphins; in effect, being groomed produces mildly narcotic effects.

The chemicals known as enkephalins and endorphins (collectively known as endogenous opiates) are produced in a region deep within the brain called the hypothalamus. They play an important role in our everyday lives as the brain's own painkillers. Their chemical signature is virtually identical to that of the more familiar opiate drugs such as opium and its derivative morphine, and they behave in rather the same way, by dampening down the pathways in the nervous system that produce pain signals. It is because morphine and other opiates are so similar in chemical structure to the endorphins that we so easily become addicted to opiate drugs. The various sites scattered throughout the brain that act as receptors for endorphins readily take up artificial opiates. However, we don't become addicted to endogenous opiates in quite the way we do to opium and morphine because the brain only produces its natural opiates in relatively small quantities. Tens of millions of years of evolution have ensured that the system only produces what it needs. Unfortunately, of course, we can flood the body with artificial opiates, thereby inducing the hypernarcotic effects associated with these drugs.

The opiate system probably evolved in order to dampen the effects of pain caused by injury, and as such it forms part of the body's carefully balanced system for handling pain. Pain is important because it warns us that something fairly drastic is happening (or is about to happen). In evolutionary terms, it has the essential function of giving us sufficient warning to allow us to remove the offending object (or ourselves) before serious damage is done. The fast pain channels of the nervous system relay the information to the brain that the skin has been broken, so that evasive action can be taken.

But once action has been taken and the danger averted, the broken skin remains broken and painful. This is where the endorphins come into play. One of their functions seems to be to dampen down the pain system to enable you to get on with the more important business of living once you have removed yourself from the dangerous situation. Were it not for the brain's opiate system, you would spend your time rolling around in agony to no great purpose. Because the opiates are released into the bloodstream, they act slowly. They take time to build up and time to dissipate, unlike the nervous system's fast-acting pain channels. It seems to be yet another example of the body's finely tuned balancing act between conflicting systems.

The endorphin system seems to respond best to the monotonous repetition of low-level stimulation. The steady pounding of jogging is just the kind of stimulus that seems to be most effective at producing endorphins. Indeed, regular joggers actually become addicted to their sport because it produces mild opiate highs for them. When they are prevented from having their daily fix, they suffer all the symptoms of opiate withdrawal: tenseness, irritability, sometimes even a mild form of the shakes.

Opiate highs of this kind are easy to induce; any kind of monotonous stress on the body produces them. Animals in captivity have long been known to pace up and down endlessly, and it has recently been shown that this stimulates opiate production – as good a way as any, I suppose, of alleviating the boredom of being caged. Workaholics probably produce the same effect, since psychological stress seems to be just as effective as physical stress. In this case, the intense concentration and high levels of brain-cell activity seem to act in much the same way as the jogger's pounding along the streets. Like joggers, workaholics suffer the classic withdrawal symptoms when prevented from working.

The body even seems to anticipate the need for opiates. Marathon runners show a marked increase in the level of opiate production a day or so before a big race. Women generate particularly high levels of opiates during the last three months of pregnancy, an invaluable preparation for the final moments of labour.

Opiates, then, play a very important role in the body's chemistry. The surprise for us is to find them turning up in the context of grooming. Research has shown that monkeys who have been groomed have higher endogenous opiate levels than those who have not. Moreover, minute doses of morphine were sufficient to suppress grooming behaviour; when the brain's opiate receptors were flooded, the monkeys were no longer interested in grooming. And when the natural production of endogenous opiates was blocked by giving the monkeys small doses of naloxone, a drug that neutralizes morphine, they were more irritable than normal and kept asking to be groomed by their cage-mates.

The mechanisms that make grooming an attractive activity seem directly related to its ability to induce a state of relaxation and mild euphoria. This, if you like, is the reinforcer that makes monkeys willing to spend so much time in what would otherwise seem a pointless activity. Even though grooming ensures that the fur is cleaned and the skin kept free of debris and scabs, the time devoted to it by species like baboons, macaques and chimpanzees far exceeds that actually needed for these simple purposes.

However, inducing opiate highs is unlikely to be the evolutionary reason why monkeys groom so much. It may be fun to space out on mutual mauling, but in a world full of predators it can be a dangerous thing to do. Opiate highs are surely the *mechanism* that encourages animals to spend so much time grooming, but something more useful is needed as the evolutionary selection pressure to drive it along. That selection pressure seems to have been something to do with cementing bonds of friendship.

It is quite common for natural selection to hijack entire motivational and behavioural systems in this way in order to use them for other purposes. One example is the way feeding and diving behaviour has become incorporated into the courtship rituals of ducks and grebes. The most remarkable example, however, is the way the three rearmost bones of the reptilian jaw were used to form the three little bones in the middle ear in the earliest mammals as they evolved out of their reptile ancestors. The reptile's lower jaw consists of five bones on either side. When the earliest mammals first evolved from reptilian ancestors, the front two bones on each side fused to form the jaw as we now see it in ourselves and all other modern mammals. The other three gradually decreased in size and became part of the hearing system. They now form a chain of tiny bones that transmits sound waves from the eardrum at the junction of the outer and middle ear to the cochlea, the organ in the inner ear that registers and encodes sounds as neural messages to the brain. In fact, this shift in use is not as surprising as it might seem, because the reptile's jaw is part of its hearing system, helping to transmit vibrations from the ground to its hearing organs. When you think about it, borrowing the jaw bones to form the ossicles was a very natural thing to do.

The capture of grooming's motivational system seems to correlate in each case with the evolution of relatively large groups and the invasion of a more terrestrial, open habitat. The increase in group size seems to be a direct response to the invasion of habitats where the animals are more exposed to the risk of predation. Arboreal forest monkeys, like the Old World colobines and all the New World monkeys, tend to live in relatively small groups. But baboons, macaques and chimpanzees are more terrestrial and prefer the more open habitats on the forest edge. Here the risk from predators is much higher, partly because it is easier for predators to approach prey undetected and partly because there are fewer trees into which the animals can escape. These species solve this problem by being larger than the average primate and, more importantly perhaps, by living in unusually large groups.

The need to live in large groups raises a host of problems, however. There are many direct costs, not the least being the need to cover a proportionately larger area each day in order to provide the same quantity of food per group member. That, of course, means you have to travel further, which in turn will expose you to greater risk of predation out on the open plains. With all that extra travel, you will burn up more energy, which in turn means you have to eat more to provide all that extra energy, which in turn means you have to travel further ... The whole process soon becomes a vicious circle. Eventually, of course, this particular circle does grind to a halt, but in the meantime you have added quite a bit extra on to the average working day for every extra body you want to have in your group.

The most serious costs are, however, the indirect ones. These come in the form of heightened levels of competition for food and sleeping sites and increased levels of harassment and stress. With so many more bodies milling around in the same fig tree, it's inevitable that there will be more competition for the best figs. The bigger thugs will get their way, and the lesser thugs will be marginalized into the less enticing patches on the outer edge of the tree. There the figs may be fewer in number, while those that are available may be of poorer quality (parasitized by wasps, nibbled at by squirrels); in addition, any position on the outer edge of the tree is more vulnerable to birds of prey such as monkey-eating eagles. The edge of the group is never a good place to be.

The press of bodies around the best and safest feeding or sleeping sites may lead to frequent trampling underfoot, both metaphorically and literally. The constant hassle of being moved on, of being harassed by those anxious to reinforce an ambivalent dominance relationship – all these add up throughout the day to a considerable strain on the nervous system of lower-ranking animals. And, of course, the lower your rank and the larger the group, the more individuals there are to harass you. Even being harassed just once a day by each member of the group can amount to a great deal of stress for the lowest-ranking animal in a group of 30-40 adult baboons.

Harassment of this kind takes its toll on the animal's bodily systems. It seems that psychological stress is every bit as effective as physical pain in stimulating opiate production. Persistent harassment can have a debilitating effect on the immune system, producing clinical depression as well as increased susceptibility to disease. But it also has an unexpected side-effect on the reproductive system, leading to temporary infertility.

It turns out that the endogenous opiates we met earlier are also implicated in the control of puberty and the menstrual cycle, though why they should be thus involved in reproduction remains obscure. Barry Keverne and his colleagues at Cambridge University have demonstrated that the stress of being low-ranking produces high levels of endogenous opiates in females, and this in turn produces infertility. It seems that the harassment meted out by other group members is as stressful as anything else and precipitates the production of endogenous opiates, which then wreak havoc on the reproductive system.

The endocrinology of this process is now quite well understood. The opiates released from the brain block the production of the hormone GNRH (gonadotrophic releasing hormone) in the hypothalamus. In the absence of the chemical kick provided by GNRH, the pituitary gland near the base of the brain does not produce leutenizing hormone (LH). LH is the hormone that stimulates the ovaries to switch over from the production of progesterone to the production of oestrogen, so triggering ovulation. Hence, with endorphins blocking the production of GNRH, the cascade of hormonal events that produces ovulation fails to occur. The result is an anovulatory menstrual cycle – a cycle that appears to be quite normal (though perhaps slightly longer than usual), but does not involve the release of an egg from the ovaries.

Just how dramatic the implications of this can be is illustrated by what happens in gelada baboons. Our field studies of wild populations of this species have shown that low-ranking females experience rather low levels of harassment: on average, about two mild threats a day from each female in the group, with only about one of these escalating into a serious attack in an entire week. Even then, these could not be considered as much more than spats on a street corner, certainly nothing remotely as traumatizing as a serious mugging. Yet these seemingly trivial levels of harassment are sufficient to result in about half an offspring less over the course of a lifetime for every rank place that a female drops. This may not sound too much, but when you realize that the maximum from which you start is only about five offspring, this represents a loss of about 10 per cent of your lifetime reproductive output for each drop in rank. By the time she is in a group of 10 females, the lowest-ranking female can expect to be functionally infertile.

Because the low-ranking females among our gelada had slightly longer menstrual cycles than high-ranking ones, we suspected that opiate-driven suppression might be the cause. At the time, we had no more than circumstantial evidence to support this hypothesis. One obvious hint was the fact that low-ranking females suffered more harassment than higher-ranking ones. However, our suspicions were later confirmed by a study of captive gelada at New York's Bronx Zoo carried out by Colleen McCann; she was able to show that low-ranking females do indeed have higher levels of circulating endogenous opiates, as well as higher frequencies of anovulatory cycles, than high-ranking females.

An even more extreme case has been studied by David Abbott at the Wisconsin Primate Center in the USA. The marmosets and tamarins are tiny South American monkeys, most of whom weigh less than half a pound. They live in family groups with a single breeding pair who are the mother and father of all the offspring in the group. The continuous low level of harassment meted out by the mother creates sufficient stress to prevent her daughters from undergoing puberty. As with the gelada, this harassment is barely noticeable to the untrained observer. But accumulated over days and weeks, it is enough to disrupt completely the natural development of the young animals. Stuck in a state of suspended reproductive animation, the daughters act as helpers-at-the-nest for their younger brothers and sisters as these are born. Eventually, when a territory becomes vacant nearby, they will leave and settle down. Once away from their parents' influence, the youngsters go through puberty straightaway, and may even conceive within a matter of weeks of being paired with a male of their own.

Perhaps we should not be too surprised by all this. After all, most of us are well aware how harrowing persistent low-level harassment can be. There does not have to be any physical contact. Indeed, that is the strange thing about it: actual physical contact somehow seems to make things better, by breaking the psychological tension. The occasional disparaging remark or glare across the room seems to be much more effective. It is the self-induced anxiety stimulated by such mild stressors that does the damage.

Even humans seem to suffer from stress-induced reproductive suppression. Two familiar examples involve career women and infertile couples. Career women in high-stress jobs (such as those in the media and finance) often find it difficult to conceive; reducing the stress levels under which they work often solves the problem. Similarly, one of the commonest stories associated with infertility concerns the couple who, having eventually given up all hope, decide to adopt; then, within a matter of months, the woman conceives. It's as though the decision to give up and adopt removes the stress of trying desperately for a child of your own. The moment the brakes are removed, the system bursts naturally into action. This is exactly what happens in the young marmosets.

I should add that this is by no means a single-sex affair. Although much less work has been done on the rather boring endocrinology of the male, it has been enough to suggest that men are by no means immune to this effect. Sperm banks (which store sperm for use in artificial insemination) obtain most of their 'donations' from medical students. They claim they can always tell when the exam period is approaching because the sperm counts (i.e. the number of live sperm per unit volume) of the samples they receive plummet dramatically as the students become more stressed. One recent study in the USA reported that males who regularly ran more than 60 miles a week had significantly reduced sperm counts compared with their less active colleagues.

The havoc that harassment and competition can play with a female's reproductive potential is clearly devastating and some mechanism is needed to defuse the problem, otherwise primate groups would rapidly collapse and disperse. There would be no advantage in females staying in groups if the result of doing so was never to be able to breed.

The solution of choice is, it seems, to form coalitions. Coalitions allow you to defuse the opposition without driving them away. After all, the whole purpose of being in a group in the first place is to permit you to survive better in an area fraught with the risks of predator attack. To drive off these other group members would re-create the very problem the groups were formed to avoid. Forming a coalition with someone else allows you to act in mutual defence, so reducing the frequency and intensity of harassment without driving your harassers away. The group itself thus achieves a state of dynamic equilibrium in which the forces of dispersion are delicately balanced by the forces of collaboration. This balancing of opposing forces is the great evolutionary achievement of the higher primates.

Servicing the relationships on which these coalitions are based is inevitably a crucial business for primates. Grooming, as we have seen, plays the key role in this. Although it is not clear exactly why it should be so effective in this respect, grooming does have a number of features that could contribute towards greater trust between coalition partners. For one thing, it is a simple statement of commitment: I'd rather be sitting here grooming with you than over there grooming with Alphonse. After all, spending as much as 10 per cent of your day grooming with someone is an immense investment of time. Whatever the intrinsic pleasures of grooming may be at the physiological level, the fact that you are prepared to make a commitment on this scale is an impressive declaration of interest and, ultimately, loyalty. Were it just a matter of eliciting opiate highs or keeping the fur clean, any old grooming partner would do. To keep returning, day after day, to the same one bespeaks a level of devotion that few other statements could match.

In addition, the very act of grooming requires you to develop a sufficient degree of trust in a grooming partner so that you are prepared to relax and let them do as they will with you. In a relaxed state, you are always open to the risk that they will exploit the opportunity to deliver a punishing attack. If you nod off, you are totally at their behest as regards warnings about approaching predators or ill-intentioned members of your own group.

The Swedish biologists Magnus Enquist and Otto Leimar have pointed out that any highly social species faces a considerable risk of being exploited by free-riders: individuals who claim a benefit at your expense on the promise to return it later in kind, but in fact fail to do so. They have shown mathematically that free-riding becomes an increasingly successful strategy as group size gets larger and the groups themselves become more dispersed.

The root of the problem is the difficulty that individuals have uncovering the free-rider's cheating ways. In large and dispersed groups, the free-rider can always keep one step ahead of discovery. By the time his erstwhile coalition partners have discovered that he won't put himself out to fulfil his obligations to them, he has moved on to form an alliance with someone else in the group. It takes time for the whispers of his unreliability to filter across from one individual to another within the group. And by then, he has moved on to a neighbouring group, and the whole process begins all over again.

Enquist and Leimar argue that this problem can be controlled by making the formation of a coalition a costly business, by individuals demanding some token of commitment before agreeing to become involved. Each time the free-rider wants to move on, he has to make a heavy investment before he can gain any advantage. Having done so, he might as well stick with the alliance, since he would be unable to form a less costly one elsewhere. Grooming, they suggest, neatly fulfils the requirements for such an investment because it costs a great deal of time. Moreover, time spent grooming with Jane is time that cannot be spent grooming with Penelope, so making it difficult for a prospective free-rider to run several equally effective alliances simultaneously.

Such tokens of good faith occur quite widely among animals. Perhaps the best studied is courtship feeding, a form of behaviour observed in birds as different as dabchicks, kingfishers and bee-eaters, as well as in insects such as scorpion flies. Among birds, for example, both the male and the female are needed to incubate the eggs or to feed the chicks once they have hatched. However, it's always possible for males to cheat, leaving the female to look after the eggs while he finds another female with whom to mate. And if the female can only raise one chick successfully out of the five or six eggs she laid, it will pay the male to abandon her if, with his help, a second female can successfully raise most of her eggs. Caught in the ultimate cruel bind, the deserted female is faced with the choice between going ahead on her own and hoping for the best, or wasting her entire investment by abandoning the eggs in order to start again with another male.

In many species, the females will only mate with males who bring them gifts, often in the form of food, during the courtship ritual. Obliging the male to make an expensive commitment before agreeing to mate with him forces him into investing just enough in the female to make it not worth his while abandoning her. The cost involved in hunting down another morsel of food – fish in one species, perhaps a field mouse in another – and then finding a second unmated female to offer it to is sufficient to make him stay where he is.

Monkey Chatter

Although grooming is the principal means monkeys use to service and reinforce their alliances, it is not the only one available to them. Monkeys are also highly vocal. The small South American marmosets and tamarins, for example, twitter away nineteen to the dozen as they forage through the tangle of branches and lianas in the forest understorey. These are largely contact calls, designed to keep the small groups together while moving through the dense Amazonian forests. Tamarin groups are typically small, often a single breeding pair with an associated adult helper or two, plus up to four offspring. Even so, the small size of these squirrel-like monkeys (an adult weighs barely a fifth the weight of a bag of sugar) combined with the dense tangle of vegetation make it easy for the animals to lose track of each other as they wander through the forest on their daily search for food. Their twittering, bird-like contact calls allow the groups to move in a co-ordinated way.

Many primate species have calls of this kind, including most of the Old World monkeys. Baboons foraging through open wood-lands keep up an intermittent susurration of grunts. For years, these were presumed to be no more than generalized contact calls. The grunts seemed like the flashing beacon of a lighthouse, letting everyone else in the group know where you are.

However, in the early 1980s the American primatologists Dorothy Cheney and Robert Seyfarth began to suspect that there might be more to the grunts of their vervets than our ears had initially suggested. So they taped the grunts made by adult monkeys, making very careful note of the circumstances under which each call had been given. By analysing the calls on a sound spectrograph – an instrument that shows the distribution of sound energy at different frequencies – they were able to demonstrate that there were very subtle, but quite consistent, differences in the sound structure of calls given under different circumstances. Calls given when approaching a dominant animal were different from those given when approaching a subordinate one, and these were different again from those given when spying another group in the distance, or when the caller was about to move out on to an area of open grassland away from the safety of trees.

They then took their recordings and played them back from hidden loudspeakers to other animals when the caller was not in view. The results were very striking. The animals hearing the calls responded in appropriate ways to the different kinds of grunt. They would look up suddenly when hearing the grunt given by an animal that was dominant over them, but ignored the calls of one that was subordinate. They would stare in the direction the loudspeaker appeared to be facing when they heard grunts recorded when the caller had spied another vervet group in the distance. And they would peer intently in the direction of the speaker itself on hearing a grunt given by someone moving out into an exposed position. Yet all these grunts sounded indistinguishable to the human ear, and had been treated as pretty much the same by previous scientists.

It seemed that, after all, a grunt was not just a grunt; a considerable amount of information was contained in the detailed structure of the call. The situation was rather akin to a native English speaker venturing for the first time into China – or if it comes to that, a native Chinese speaker venturing into England. Suddenly you are surrounded by babble, a cacophony of sound with little or no definable structure. You understand nothing. You cannot even identify separate words. It might just as well be gibberish. Yet what these people around you are doing is clearly rational and involves communication. Later, with infinite patience and a great deal of practice, you learn to pick out words, then phrases and finally whole sentences. And, behold! the whole business is in fact a masterpiece of communication. Immensely complex concepts were being transmitted from speaker to listener. Sophisticated metaphysical arguments were being debated. Poetry of exquisite beauty was being recited and discussed.

So the question arises: was our view of vervet vocalizations simply a reflection of our ignorance of vervetese? Are we like native Chinese speakers suddenly transported to England when we stand beside a vervet group on the Africa savannahs? After all, it takes a child a decade or so to develop fully fledged Chinese or English. Even if the native Chinese speaker works hard at it, the English he speaks after several years' practice is still halting and emasculated, his childlike mistakes the subject of much amusement. Have we been making a serious mistake then in assuming that, after a mere month or two in the field, we know all there is to know about vervet grunts? A vervet, after all, has spent at least five years in the equivalent of childhood listening to the calls of the adults in its group, picking up nuances here, minor inflections there. Just what *does* a vervet hear when it hears a grunt?

One thing has become abundantly clear from the past decade's research: primate communication is much more complicated than anyone had previously imagined. Vervets, for example, clearly distinguish between different types of predators and use different calls to identify them. They distinguish ground predators like leopards from aerial predators like eagles, and both of these from creepy-crawlies like snakes. Each type of predator elicits a different type of call. In a classic series of experiments, Cheney and Seyfarth showed that vervets responded appropriately to the disembodied calls when they were played back from a hidden loudspeaker. The monkeys ran for the trees on hearing the leopard call, dived out of the tree canopy on hearing the eagle call, and stood on their hindlegs and peered into the grass around them on hearing the snake call. They did not need to see the predator itself. Nor did they need to see what the caller was getting excited about, or how excited the caller actually was. All the information they needed to identify the type of predator was contained in the sound itself – just as all the information you need to identify a leopard is contained in the sound 'leopard', but not in the sounds 'watch out!' or 'help!'

Other species of monkeys and apes exhibit similar complexities in their vocal exchanges. The small family units of the gelada are very strongly bonded, and the members keep up a constant chatter among themselves. These exchanges involve complexly structured whines, moans and whinnies expressed with all the intonations of speech. Listening to them with eyes closed is like sitting at the far end of a restaurant or a bar where you can hear the rise and fall of speech, the alternation of voices as the speakers take it in turns, but you cannot make out the words themselves.

While feeding, gelada use vocalizations in complex exchanges that seem designed to allow individuals to maintain contact with their favourite grooming partners while they are apart. In addition to these conversation-like exchanges between friends, they use moans and grunts as prompts when grooming. All too often, the recipient of grooming will become so relaxed that it falls asleep. Eventually, when the groomer tires of grooming and wants to be groomed in its turn, it will stop and present a raised shoulder or arm to its companion. In the normal course of events, the companion will immediately turn and begin to groom the proffered body part. But dozing in the warm sunshine, it may sometimes be unaware that grooming has ceased. Its companion will often give a quiet grunt, as much as to say 'Hey! It's your turn!'

When a female gives birth, the new baby soon becomes a subject of fascination to the other young females in the group, especially those just past puberty that have not yet given birth themselves. When one of these young females approaches an older sister, or her mother, who has a new baby, the excitement in her voice is palpable. Her contact calls race up and down the vocal scale like an excited child's, the words tumbling over themselves as they pour out in a confused torrent. The emotional overtones are clear in both cases, but this is more than simply mindless gibbering: here is real communication of the excitement of the moment.

These new findings are raising doubts about the conventional wisdom that only humans have language. Linguists and psychologists have always insisted that only our species has true language. Of course, other animals communicate with each other, and indeed with us – as does your dog or cat when it wants to be let out or go for a walk. But, insist the linguists, what these animals do is mere communication. It does not rank as language because they cannot express abstract concepts by their barks and grunts. Indeed, many of them maintained (and still maintain) that all communication by non-human species is largely confined to the expression of emotional states. Dogs bark when excited because that's the kind of sound their respiratory system produces when they get excited.

In the 1960s the linguist Charles Hockett proposed a set of some eighteen features which could be taken to define true language. The four most important of these were that genuine verbal language is (1) referential (the sounds refer to objects in the environment); (2) syntactical (it has grammatical structure); (3) non-iconic (words do not resemble the objects they refer to – unlike the 'word' *moo*, for example, which is clearly intended to resemble the sound made by cows); and (4) learned (as opposed to being instinctive).

These criteria were designed to make clear the difference between true language and, say, the 'language' of honey bees. During the 1950s the ethologist Karl von Frisch had demonstrated that honey bees communicate about the location of good sources of nectar when they return to the hive after a foraging trip. Von Frisch discovered that a returning forager will often execute a rather stereotyped figure-of-eight dance on the vertical surface of the honeycomb. Through a combination of ingenious experiments and very careful observation, he was able to show that the speed of the dance indicated the distance of the nectar source from the hive, while the angle of the bar of the eight to the vertical indicates the compass direction relative to the sun.

Everyone was impressed by the bees' remarkable achievement, and rightly so. It is surely one of the miracles of the natural world. These dramatic findings certainly made linguists and philosophers sit up and think. Was this true language? Because if so, humans were no longer so unique. To some it raised the serious prospect that, Dr

Doolittle-like, we might be able to learn to speak to animals.

Once the flurry of excitement died down, however, it soon became apparent that what the bees did should not really be counted as language in the human sense. It was very stylized, and could communicate only a limited number of facts about an extremely restricted range of topics. It also appeared to be instinctive; it was very doubtful whether the bees 'knew' what they were saying.

In the meantime, however, the suggestion that animals might have primitive forms of language had led to serious attempts to teach them languages. For obvious reasons, perhaps, the early attempts all focused on our nearest relatives in the animal kingdom, the chimpanzees.

Ape-Speak

During the 1950s there were two celebrated attempts to teach chimpanzees to speak English. The Kelloggs and the Hayeses both tried raising a baby chimpanzee along with their own new-born infant, giving both infants equal attention and the same opportunities to learn to speak. The results proved disappointing. Vicki, the young chimp reared by the Hayes family, managed to produce no more than half a dozen words, and then barely audibly. Worse still, so far from the chimp learning human habits, as they had hoped, the Hayes' own child picked up innumerable bad habits from the chimpanzee. Because chimps mature much faster than human infants and quickly get into mischief, Vicki had proved a wonderfully attractive role model for Hayes junior. The Hayeses gave up.

At the same time, it began to dawn on everyone that chimps would never learn to speak, because they lacked the vocal apparatus to produce the sounds necessary for human language. This requires a deep-set larynx which provides a large resonating chamber at the back of the nose and mouth, as well as careful control of the vocal chords to act as vibrators. Chimps, it seems, do not have the crucial anatomical devices.

This recognition led to an alternative approach in the 1960s. Another husband and wife team, the late Trixie Gardner and her husband Alan, began a new study with a young female chimp named Washoe. This time, Washoe was taught a sign language rather than a vocal language. Accepting that chimps would never speak, the Gardners argued that they might learn a sign language, because gestures were a natural part of the everyday communication of wild chimpanzees. The Gardners chose the American deaf-and-dumb language ASL, which uses gestures to stand for concepts or words. So Washoe was brought up in an environment where everyone who came into contact with her always communicated in ASL.

Washoe proved to be the wonder of the age. Eventually, she learned a hundred or so signs. But not everyone was convinced. Several psychologists and linguists argued that Washoe had demonstrated nothing more than the ability to copy her human carers. They pointed to the fact that many of her signings were repetitious, and often seemed to require prompts from the humans. She rarely produced 'sentences' that were more than two signs long (when all the repetitions were excluded). The supposedly novel sign combinations she produced – she was said to have spontaneously produced the signs 'water' and 'bird' the first time she saw a swan – were no more than chance combinations. Washoe's communicativeness, the critics insisted, lay only in the imaginative interpretations of the Gardners.

The Gardners spent the better part of the next twenty years trying to test Washoe in ever more rigorous conditions to prove that she could use ASL as a language, while the Gardners' critics devised ever more ingenious attempts to demonstrate that everything Washoe had achieved could be explained as a 'Clever Hans' phenomenon.[1]

In the meantime, the 1970s saw the start of several new projects. Two of these also involved ASL (the gorilla Koko and the orangutan Chantek), but two others tried to get around the criticisms

---

1. Clever Hans was a circus horse who, in the early years of this century, excited much interest by apparently being able to add two numbers together and then count out the answer by tapping with a hoof. Careful study revealed that his owner was inadvertently giving Hans cues to stop tapping: Hans was detecting the man's soft indrawing of breath when he got to the right number.

levelled at the Gardners by teaching chimpanzees pictorial languages. The psychologist David Premack taught a chimp named Sarah and several of her cage-mates a language that used coloured plastic shapes to stand for words (or concepts). The plastic shapes were magnetized so that they could be assembled on a metal board to form sentences. The other study, initiated by Duane Rumbaugh, involved two chimps named Austin and Sherman who were taught a computer keyboard language named 'Yerkish' in honour of one of the fathers of modern comparative psychology, Robert Yerkes. The keyboard consisted of a large array of coloured shapes rather than letters, with each key representing a word. Later, Sue Savage-Rumbaugh taught the same language to a bonobo named Kanzi. Kanzi was to become the Einstein and Shakespeare of the chimpanzee world rolled into one.

The debate about language, however, rumbled on. Irrespective of how clever Sarah and Kanzi might be at answering questions or following instructions, were they really *using* language in the sense that a human child does? Did they understand *grammar*? Did they understand abstract relations like 'is bigger than'?

I think it fair to say this research has convincingly demonstrated that chimps understand several important concepts, including numbers, how to add and subtract, the nature of basic relations (such as 'is bigger than', 'is the same as' and 'is on top of'), how to ask for specific objects (mostly foods) or activities (a walk in the woods or a game of chase), and how to carry out complex instructions ('take the can from the fridge and put it in the next room'). Kanzi can translate readily from one modality to another: for example, by pointing to the correct keyboard symbols for spoken English words heard through a set of headphones. This is thought to be an especially crucial prerequisite for language because it underlies our ability to translate from hearing to speaking (and, of course, writing).

However, despite all the effort devoted over the past three decades to training apes to use language, none of them has progressed convincingly beyond the simple two and three-word sentences typical of two-year-old human children. Kanzi's extraordinary understanding of language notwithstanding, his abilities are largely limited to asking for things he wants, carrying out instructions and giving the correct one-word answer to logically complex questions. He does not engage in the kind of spontaneous, apparently effortless chatter of the two-year-old human child learning to speak. At that stage, human children spend a great deal of time simply naming objects, apparently for the hell of it. 'Look, mummy, car!' ... 'Yes, dear, *another* car ...'

Even in contrast with the fluent witterings of the gelada's contact calling, the conversational skills of language-trained apes seem stilted and laboured. Chimps may, at best, have a foot on the language ladder, but it is not all that we might have expected of an ape standing on the very threshold of humanity. How is it, then, that one species of ape made that transition? To answer this question, we need to understand what human languages are used for and why they evolved.

CHAPTER 4

Of Brains and Groups and Evolution

By common consent, monkeys and apes are the intellectual geniuses of the animal world. Their antics and imitativeness, their bright eyes and mischievousness, tell it all. Since time immemorial, humans have recognized their special human-like qualities. Inevitably, we are led to ask why it is that monkeys and apes are so intelligent.

Why Do Monkeys Have Big Brains?

Defining intelligence has always been a thorny problem, and psychologists have got themselves into deep trouble over the years trying to come up with sensible ways of measuring it. There are two problems: one is trying to say what intelligence *is*, and the other is trying to find a reliable measure of it.

Intuitively at least, the problem of definition is not too difficult. Most of us would probably agree that being intelligent means being able to solve problems that other people can't. Einstein was intelligent because he thought up the Theory of Relativity, and there are rather few of us who can either understand what it means or follow the mathematical arguments he used to develop it. Psychologists have, of course, recognized that there may be different kinds of intelligence: social intelligence is not quite the same thing as being clever at science, and that in turn is not quite the same as the qualities that characterize a good novelist or musician. Nonetheless, there remains a firm belief in some underlying common factor that can be applied in different domains. Psychologists in the early part of this century referred to this as the 'G factor' (for 'general intelligence'), while at the same time hedging their bets and identifying a set of components to IQ that reflect skill at visual problem-solving, verbal and arithmetical skills, and logical reasoning.

A more serious problem has always been the likelihood that any measures we use might simply measure an individual's motivation to answer questions or his general knowledge, rather than his native intelligence. It was a surprisingly long time before psychologists realized that the reason black Americans did so badly in conventional intelligence tests was not so much because they were less intelligent than white children, but simply because they were less widely read. Tests which assumed that everyone had had the same opportunities to learn about geometry inevitably measured only educational experience, not native wit.

Nonetheless, the general principle that there is something we can label as intelligence which varies between individuals and between species remains widely accepted. Our problem has more to do with designing tests that allow us to measure this elusive quality in different species in ways that are fair to everyone. One solution has been to measure brain size rather than performance in practical tests.

One of the conventional wisdoms of popular culture has always been that bigger is better: animals with bigger brains must be more intelligent. After all, humans have bigger brains than, say, dogs. But there were some inconvenient anomalies: elephants and whales have bigger brains than humans, so does this mean these species are more intelligent than us? In the early 1970s the psychologist Harry Jerison realised that absolute size isn't everything. You would expect a large animal to have a large brain because it has a great deal more muscle mass, not to mention the size of the body's other organs, to keep going. More leg muscles require more instructions from the brain in order to keep them co-ordinated in the right way.

It was not total brain size we should be interested in, he argued, but *relative* brain size. Intelligence is a consequence of how much spare computer capacity you have left over after you have taken out everything necessary to keep the body ticking over and working properly. To try to get at this, Jerison suggested that we first plot brain size against body weight. This would give us a general relationship between brain size and body size that, more or less, measured the amount of brain tissue required for basic bodily functions. Whatever was left over was the spare capacity available for clever things like problem-solving.

With this in mind, Jerison plotted brain size against body weight for all the species he could find, everything from dinosaurs to primates. The results were rather enlightening (see Figure 1). The distributions revealed that whole groups of animals lay on higher planes than others. The species values for dinosaurs and fishes lay below those for birds, while those for mammals lay above the birds. More interestingly still, the values for primates lay above those for other mammals, and similar distinctions could

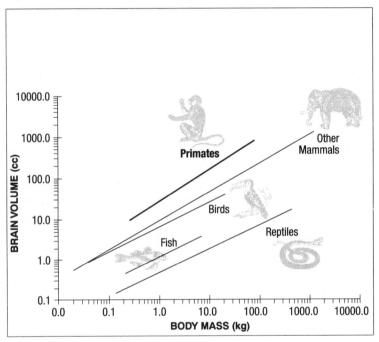

Figure 1. When brain volume is plotted against body weight (or mass) for various groups of animals, most of the points for each group lie along separate lines. Primates have larger brains for a given body size than all other species, while reptiles have the smallest. Note that both axes are plotted on a logarithmic scale: the relationship between brain size and body size is in fact curved, indicating that brain size does not grow at quite the same rate as body size. Plotting the data for a curvilinear relationship in log form produces a straight line, which allows the patterns to be seen more clearly.

be made even within the mammals as a whole. At the bottom of the mammal pile came the marsupials (kangaroos and their allies); above them lay the insectivores (shrews, hedgehogs), then the ungulates (sheep, cattle, deer and antelope), followed by the carnivores (cats, dogs, racoons, etc), and finally the primates at the highest level. Even within the primates, prosimians (the lemurs of Madagascar and the galagos of mainland Africa) seemed to lie on a lower scale, with smaller brains for body size, than the advanced primates (monkeys and apes).

Measured in this way, we humans have a brain about nine times larger for body size than is usual in mammals in general. Our brain is about 1600 cc in volume, but a typical mammal of our body weight (55 kg) would have a brain of only 180 cc. Even more impressive is the fact that we have a brain about twelve times larger than one would predict from the figures for insectivores. This is the group of small animals such as shrews, hedgehogs and moles that are generally regarded as being most representative of the kinds of primitive mammals that lived some 60 million years ago, and from which all living mammals derived.

This raises a fundamental question: why do some species have bigger brains than others? Why do primates, in particular, have bigger brains than mammals such as cats and dogs? Moreover, why do some primates (such as chimpanzees and humans) have bigger brains for body size than other primates (like the lemurs of Madagascar and the leaf monkeys of Asia)?

During the first half of this century psychologists tended to view intelligence as abstract reasoning ability. The fact that one animal was more intelligent than another aroused no particular comment. That was simply the way the world was. But from a biological point of view, it just doesn't make sense. Brain tissue is extremely expensive to grow and to maintain. Your brain accounts for only 2 per cent of your body weight, yet consumes around 20 per cent of all the energy you take in as food. In fact, brain tissue is so costly that it cannot be there by chance. The fact that an organism has a large brain means that it really must need it very badly, otherwise the forces of natural selection will inexorably favour individuals with smaller brains simply because they are cheaper to produce. Animals that have to spend a lot more time feeding to provide the fuel for their big brains (or mothers that have to do even more to provide for their infant's brain as well as their own) expose themselves to proportionately greater risks of predator attack, not to mention starvation when things get tough in famine years. While their small-brained compatriots can retreat into safe nooks and crannies, the luckless brainy ones are out there with their attention focused on the business of finding food. It's a classic trap. If you get on with the business of feeding, you don't see the predator creeping up on you; but if you keep stopping to check for predators, you have to spend longer feeding, and so increase the likelihood that a predator will eventually turn up. Something very significant must be acting to preserve so expensive an organ.

One answer suggested in the 1970s was that brains were needed to solve the problems of life and survival. Some animals needed bigger brains than others because the problems of daily living they encountered were more complex. So it is that fruit-eating animals like monkeys need bigger brains for body size than leaf-eating animals like sheep and cattle. Fruits, so the argument runs, are very patchy in their distribution, being here today and gone tomorrow. Leaves, in contrast, are there in plenty pretty much all of the time. Even if they sometimes get a bit brown around the edges, there is still enough of them to keep you going. Consequently, a fruit-eater needs a bigger brain to keep track of where these patchily distributed resources are than a leaf-eater whose foods tend to be more evenly distributed and more widespread.

There is growing evidence to suggest that the initial impetus for the evolution of super-large brains in primates (as compared to other mammals) did in fact have something to do with colour vision: finding fruits against a background of leaves is greatly facilitated by colour vision in both birds and primates. The colour-vision system found in primates is superior to that found in other mammals, and inevitably requires a great deal more computer-power. However, while a shift to a more fruit-based diet explains rather neatly why primates should have larger brains than other mammals, it does not explain why some fruit-eating primates should have bigger brains than others.

An alternative solution proposed in the late 1980s was that the unusually large brains of primates had something to do with their particularly complex social behaviour. The suggestion that social complexity might be at the heart of primate intelligence had in fact been noted by several ethologists during the previous three decades, but no one took their suggestions especially seriously until, in 1988, two British psychologists, Dick Byrne and Andrew Whiten, proposed what has become known as the Machiavellian Intelligence hypothesis.

Byrne and Whiten argued that what makes primate social groups quite different from those of other species is the fact that monkeys and apes are able to use very sophisticated forms of social knowledge about each other. They use this knowledge about how others behave to predict how they might behave in the future, and then use these predictions to structure their relationships. Other animals, they suggested, lacked this capacity and instead made do with simpler rules for organizing their social lives. Monkeys and apes, for example, were able to work out the implications that Jim's relationship with John might have for the relationship between John and themselves; recognizing that Jim was John's friend, they would know it wasn't worth asking Jim to help them against John. Other animals, by contrast, would only understand the relationships they had with Jim and John, and so might make the mistake of seeking Jim's help against John.

These two hypotheses remained somewhat at loggerheads during the late 1980s and early 1990s. Many complained that the Machiavellian Intelligence hypothesis was too nebulous to test, that there was no concrete evidence to support it, and that in any case there was plenty of evidence to support the ecological hypothesis. It was known, for example, that fruit-eating primates had larger brains than leaf-eating primates, and that primates with large brains had much bigger territories than primates with small brains.

It seemed to me, contemplating this problem from the vantage point of the 1990s, that the earlier analyses had confounded a number of different factors. Fruit-eating primates invariably have larger territories than leaf-eaters, for the very good reason that fruits are more patchily and widely dispersed than leaves. However, at least some fruit-eating primates (such as baboons and chimpanzees) are physically bigger than any of the leaf-eating monkeys, and commonly live in larger groups. The larger species thus tend to be fruit-eating and have larger territories, as well as having bigger brains and living in bigger groups. This makes the causal relationships involved difficult to disentangle. It could be, for example, that these species have large territories because they are fruit-eaters, and are fruit-eaters because they have large brains, which they need in order to hold large groups together.

Since the four variables – brain size, body size, territory size and fruit-eating – were thoroughly confounded, it was impossible to be sure that the correlation between any two of them was not simply a consequence of the fact that both had been correlated for quite distinct reasons with the third. Some way of testing between the various hypotheses was needed, to see which one correlated best with changes in brain size in primates independently of the others.

One point, however, seemed important. All the previous analyses had looked at total brain size. Yet when we look at the story of primate evolution, it isn't the whole brain that progressively increases in size as we make our way from the smallest, most primitive species like lemurs to the larger, more advanced forms like humans. In addition, psychologists have been coming to the conclusion that the mind doesn't work like an all-purpose computer that can use any of its bits to do any jobs. Rather, their experimental studies were beginning to suggest that the mind consists of a number of separate modules, each designed to do a particular task. Each of those modules might well be associated with different parts of the brain – just as vision is located in one part, language in another, and motor control in yet another.

The mammalian brain seems to consist of three main sections: the primitive brain we have inherited more or less intact from our distant reptile-like ancestors; the mid-brain and other sub-cortical areas that are mainly concerned with sensory integration and the machinery of living; and finally the cortex, the outer layer that is pretty much unique to the mammals. Within this broad arrangement, however, primate brains strike us as unusual: one particular part, the neocortex, has changed out of all proportion to what it looks like in other mammals. The neocortex is what you might call the 'thinking' part of the brain, the place where conscious thought takes place. It is a rather thin layer, being a mere five or six nerve-cells (about three millimetres) deep.

This thin slab of neural tissue is wrapped around the inner mammalian brain, and receives and sends neurones to many parts of the brain. In most mammals, the neocortex accounts for 30 to 40 per cent of total brain volume. But in primates it varies from a low of 50 per cent among some prosimians to 80 per cent of total brain volume in humans. It seemed to me that we really should be looking at the neocortex, not the whole brain. Not only was this the part of the brain that had expanded dramatically in primates (and especially ourselves), but it was also the part where the activities we associate with intelligence – thinking and reasoning – seemed to go on.

When I looked at the primates, it turned out that there was no correlation between the size of the neocortices of species and any of the obvious ecological criteria: things like the percentage of fruit in the diet, the size of the territory, the distance travelled each day while foraging, or the complexity of the diet measured in terms of the work the animals had to do to extract edible food from the substrate in which it is embedded. But group size, the one measure of social complexity available to me at the time, *did* correlate with neocortex size, and did so with a remarkably good fit (see Figure 2). All these relationships held up even when the effects of differences in body size were statistically removed in order to overcome the problem of confounded variables we noted above.

I used group size as my measure of social complexity for two reasons. Firstly, it is one of the few things that field-workers invariably count and give solid numerical values for. It was therefore easy to obtain data on group size for a large number of primate species. Qualitative observations (like saying that one species is more complex than another) aren't very helpful when it comes to testing hypotheses rigorously, because it is easy for us to deceive ourselves into seeing what we want to see. *I* may think that chimpanzees are socially more complex than baboons, but

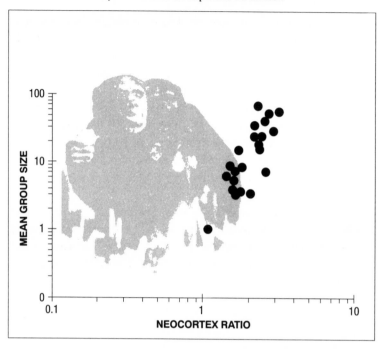

Figure 2. Mean group size for different genera of anthropoid primates (monkeys and apes) plotted against a measure of the relative size of their neocortex. The measure of neocortex size used here is the ratio of neocortex volume to the volume of the rest of the brain: this helps to correct for differences in neocortex size that are simply a consequence of differences in body size. The two axes are plotted on a log scale in order to represent the relationship between the two variables as a straight line.

you may disagree. I needed some measure of complexity that everyone could count and agree on.

The other reason was that, in one very important respect, social complexity does increase with group size. Given that primate social life is characterized by the ability of the animals to recognize relationships between third parties – Jim's relationship with John as well as John's relationship with me – there is a real sense in which the social complexity of a group rises exponentially as its physical size increases. In a group of five individuals, I have to keep track of a set of four relationships between myself and the other group members, but I have to monitor six additional relationships involving the other four individuals. In a group of twenty, I have to keep track of nineteen relationships between myself and fellow group members, and 171 third-party relationships involving the other nineteen members of the group. While my relationships with everyone else have increased roughly fivefold with the fivefold increase in group size, the number of third-party relationships I have to keep track of has increased almost thirtyfold. Crude though it may be, group size does provide one index of the amount of information-processing a social animal has to engage in.

These analyses provided striking evidence to support the Machiavellian Intelligence hypothesis. The evolutionary pressure selecting for large brain size and super-intelligence in primates did seem to have something to do with the need to weld large groups together.

The Picture Gets More Complex

When I first discovered this relationship between group size and neocortex size, I assumed it was something unique to the primates. A number of my colleagues were inclined to argue that they would believe the relationship was real *if* I could show it was also true for at least one other group of animals. Coming up with the relationship for the primates might just have been a lucky accident; showing that it held true for some other group as well would make 'lucky accident' a much less plausible explanation.

My response to this challenge was that the same relationship would only hold true in other groups of mammals if their social systems were organised in ways similar to those of primates. In other words, only if the social life of the other species also involved the formation of complex coalitions based on sophisticated manipulation of social knowledge would you expect to see the same neocortex/group size relationship. The brains of other species may well have evolved to solve the ecological problems of daily survival, but the point of the Machiavellian Intelligence hypothesis was that something extra was added during the course of primate evolution. Since everyone knew that these kinds of complex social groups are unique to primates, I was confident that only primates would show this relationship between group size and neocortex size.

But science is full of surprises. The first intimation that I might be wrong came in a letter from my colleague Rob Barton at Durham University. He had discovered that bats which live in stable social groups have larger neocortices than those that live in unstable groups. The most intriguing case in this sample was the much-maligned vampire bat. Field studies of the behaviour and ecology of these bats carried out by American biologist Gerry Wilkinson, amongst other bat biologists, have revealed that vampires are highly social animals, even though they tend to forage on their own. Once back at the roost, they seem to behave rather like miniature primates. They spend a great deal of time grooming each other. They have special friends with whom most of this grooming is done. When one of them has a bad night and fails to find an animal from which it can suck blood – licking is really what they do – its friends will regurgitate part of their own dinner for it. Days later, it will repay that debt when its friend has a bad day. Here we have a highly social species which forms tight little coalitions that act reciprocally to help each other out, that cements those relationships with social grooming. This sounds just like monkeys and apes. So it was doubly intriguing to discover that the vampire bat had by far the largest neocortex of all the bats.

This set us thinking, and we began to search for brain-size data on other groups of mammals. This was not easy to come by. Indeed, the only reason we had such good data for primates was because one particular laboratory in Germany (under the direction of the anatomist Heinz Stephan) had made a virtue out of the laborious business of slicing up the brains of primates and measuring the areas of each part, thin sliver by thin sliver. I shudder to think of the tedium involved, but it paid off in the end by providing us with a unique database. Even then, the samples were small, often just one or two specimens for each species, and covering only about 70 of the 200-odd species of living primates. (Stephan was dependent on specimens sent to him by zoos after animals in their collections had died of natural causes.)

We did, however, manage to find sufficient data for the carnivores. The carnivores looked promising because a number of the larger species, such as lions, wolves and hunting dogs, are decidedly social, and their diurnal habits make them easier for biologists to study. We were surprised and bemused to find that the carnivores slotted rather neatly on to the front end of the neocortex/group size graph for primates. Socially, carnivores were, you might say, just small-brained primates.

This finding was satisfying because it suggested there might be an underlying unity to the social lives of all mammals. We would not necessarily have to plead a special case for the primates. Instead, we could see them as a natural outgrowth from the basic mammal stock. Biologists always feel happier when they can argue for continuities of this kind. Special cases that require them to argue for large, unique mutations always make them feel uncomfortable, simply because large mutations of that kind are statistically implausible given the way our genetics actually work.

These findings raise another interesting question: just *how* does neocortex size relate to group size? All our analyses so far had been built on the assumption that the problem each animal has to face is keeping track of the constantly changing social world of which it is a part. It needs to know who is in and who is out, who is friends with whom, and who is the best ally of the day. In the social turmoil, these things were in a permanent state of flux, changing almost day by day. The animal has to keep track of all this, constantly updating its social map with each day's new observations.

But there are other possibilities. One is that the relationship between neocortex size and group size actually has more to do with the *quality* of the relationships involved rather than their quantity. This much is implied by the Machiavellian hypothesis itself, which suggested that the key to understanding brain size evolution in primates lies in the use primates make of their knowledge of other animals.

Two interpretations seem possible. One derives from the fact that coalitions form so important a part of primate social life. This suggests that the relationship between group size and brain size might in reality be between brain size and the size of the coalitions that the animals habitually maintain. The correlation with total group size would thus simply be a by-product of the fact that animals need progressively bigger coalitions to maintain the stability of larger groups.

The other derives more directly from the Machiavellian hypothesis and suggests that big groups come about not simply because the animals can manage to keep track of so many second and third-party relationships, but because they have to be far more sophisticated in maintaining the balance between so many conflicting interests. Their problem, in other words, was not just keeping track of the relationship between Jim and John and themselves, but managing a balance between the three relationships involved. Keeping both Jim and John happy at the same time is a much more difficult feat than simply remembering whether they are friends with each other.

I began discussing these ideas with the young Japanese primatologist Hiroko Kudo, who was on a visit to my research group in London. Together with Sam Bloom, one of my graduate students, we began to gather data on coalition sizes in different primate species. Our first problem was how to define a coalition, never mind counting the number of animals involved. In real life, coalitions are usually only evident for a matter of moments at the time they come into play. A typical instance goes something like this: one monkey is challenged by another – its coalition partner immediately rushes to its aid – the challenger immediately recognizes that the odds are against it and retreats – everything settles down again. Had it not been for the snarls and screams, the human observer might have missed it altogether. Clearly, we wouldn't get very far if we tried scouring the literature for these kinds of incident. We would, of course, find plenty of descriptions of them – but how could we find out the relative frequencies with which different species exhibited such behaviour from the kinds of qualitative description scientists published in books and articles?

There was, however, another possibility. As we saw in Chapter 3, primates spend hours each day ruffling through each other's fur, removing bits of loose skin or burrs caught in the fine hairs.

It's an activity they find both pleasurable and relaxing. It has been equally clear, as we saw in Chapter 2, that these extended bouts of mutual mauling are not doled out at random. Only animals that have long-standing relationships groom each other regularly. It is an activity for friends, not for acquaintances. Grooming between acquaintances is often brief, perfunctory and decidedly lacking in enthusiasm.

Other kinds of evidence reinforced the suggestion that grooming is a good guide to coalitions. Robert Seyfarth and Dorothy Cheney had shown in their experiments with wild vervet monkeys that individual monkeys are much more likely to pay attention to the distress calls of individuals with whom they have recently groomed. In addition, I had been able to show from my own studies of wild gelada baboons that animals which groomed each other were more likely to support each other in fights against a third party than were animals which rarely groomed together.

Grooming, then, seemed the obvious thing to look at. It seemed to function as the cement which welded the relationships making up a coalition. More importantly for our purposes, it is an activity that is both easy to see and common in primates, so most fieldworkers record its occurrence and use it to analyse the pattern of friendships within their study groups. So we combed the literature for studies that provided data on grooming frequencies between all the adult members of the group.

We managed to extract data for about two dozen species in all. Having calculated mean grooming clique sizes – the number of animals that form an integrated subset linked by high grooming frequencies – we then plotted them against neocortex size and total group size. Just as we expected, the mean size of grooming cliques correlated well with both the other measures.

These results suggested that, as group size increases, so the animals have to form larger and larger grooming cliques to protect themselves from the harassment they inevitably experience when living in large groups. The principal purpose for which so much computing power was needed was, it seemed, to enable primates to form long-lasting, tightly bonded coalitions.

The details of how it all worked were still far from clear, but the broad patterns were beginning to emerge. The Machiavellian theory appeared to be right. At this point, an obvious question sprang to mind.

Where Do Humans Fit In?

Given this relationship between neocortex size and group size in primates, what does it tell us about humans? We are primates, just like all the others. Indeed, the most recent evidence from molecular biology suggests that, on the basis of similarities in genetic material, chimpanzees and humans are more closely related to each other than either is to the gorilla, their next closest relative among the primates. Since this relationship between brain size and group size seems to fit chimps neatly, we should expect the same of humans too.

So what size of group would we predict for humans? Humans have a neocortex ratio of 4:1, and if we plug this value into the graph shown in Figure 2, we can read off a predicted group size for humans. The answer turns out to be groups of about 150.

Now, one's first reaction to this is disbelief. After all, humans live in cities like Tokyo and London, New York and Calcutta, places where 10 million or more people live crowded together. How can a figure as small as 150 possibly be correct?

But remember what kind of group the relationship in Figure 2 was based on. Primates live in small groups where everyone knows everyone else, at least by sight even if they don't know them from personal interaction. Human ingenuity may have produced these giant conurbations, but this does not mean that all the people who live in them are social intimates. The vast majority of the people in Tokyo and New York are born, live their lives and die without even being aware of each other.

Modern human societies are precisely that – modern. Cities appeared for the first time around 3000 years ago. Yet modern humans appeared around 150,000 years ago. People who were recognizably members of our own species, *Homo sapiens*, first appeared as long as 400,000 years ago. Throughout that long time-span, right up until the appearance of agriculture a mere

10,000 years ago, we lived as hunter-gatherers in small bands wandering through the woodlands in search of game. Perhaps the place to look first, then, is among the societies of those modern humans who still follow a traditional hunter-gatherer way of life. Dozens of such peoples can still be found today in the deserts of southern Africa and Australia and in the forests of South America.

These small-scale societies have been studied in considerable detail by anthropologists over the past century. It is, of course, true that almost all such societies have been affected by both modern Western and indigenous civilizations as they have come into contact with them over the course of the last two centuries. But even though their beliefs about the world may have changed, and the gun and the snow-scooter may have replaced the traditional bow and dog-sled, they nonetheless preserve much of the traditional way of life that has characterized them since long before written history began. In particular, the grouping patterns they adopt may not be all that different from those they had in the distant past.

All such societies are characterized by tiers of social organization that are increasingly inclusive. At the bottom of the hierarchy are the temporary overnight camps of 30–35 people, some five or six families in all. These are essentially ecological groupings, the temporary coming together of a variable number of families who find it convenient to pool resources for a while and collaborate in hunting and food-finding. At the top of the scale is the largest grouping, the tribe, typically numbering some 1500–2000 people. The tribe is a linguistic group, the set of people who speak the same language (or in the case of the more widely spoken languages, the same dialect).

Between these two levels can sometimes be found the groupings of about 500 people commonly referred to as 'mega-bands'. Below them, smaller groupings can sometimes be discerned, often referred to as clans. These turn out to average almost exactly 150 and to be very much less variable in size than any of the other kinds of groupings mentioned above.

Clans are particularly interesting from our point of view because they are often associated with some ritual function.

Among the Australian Aboriginals, for example, clans gather together once a year to organize rites of passage for the young males, to contract marriages, and generally to reinforce the sense of collective identity by rehearsing the old rituals and recounting the age-old myths and stories that tell of the people's ancestry and their relationships with the spirit world. Clan members know how they relate to each other. They can specify exactly whose great-grandmother was whose great-great-aunt, whose cousin is whose great-niece.

Could these clans be the groups we are looking for? They are certainly the right size. Moreover, they have just the kind of social properties we would expect from the groups that characterize monkeys and apes. They appear to be the largest grouping in which everyone knows everyone else, in which they know not simply who is who but also how each one is related to the others.

In fact, 150 turns out to have even more interesting properties. It is, for example, roughly the number of living descendants (including all the wives, husbands and children) that you would expect an ancestral couple to have produced after four generations at the kind of birth-rates conventionally observed in hunter-gatherer and peasant societies. What's interesting about this is that a pedigree, or family tree, five generations deep takes you back to grandma's grandma, or as far back as any living member of the group can remember as a matter of personal experience. In other words, as far back as anyone can vouch for particular relationships: only within the circle of individuals defined by those relationships can you specify who is whose cousin, and who is merely an acquaintance.

But it is one thing to show that these groupings exist in the small-scale societies of hunter-gatherers. It is quite another to suggest that they are characteristic of *all* human societies. What about the technologically more advanced societies, the societies of agricultural peoples and our own large-scale industrial economies? It turns out that most societies have some kind of grouping of about this size. On the basis of the number of dwellings present, archaeologists have suggested that the villages of the earliest farmers of the Near East around 5000 BC typically numbered 150 people. The villages of modern horticulturalists in Indonesia and the Philippines, as well as those in South America, are also typically around 150 in size.

Nor is it just non-European agriculturalists that group in this way. The Hutterites have pursued a fundamentalist Christian way of life based on communal farms for nearly four centuries, originally in Europe but now in Dakota and southern Canada. Though each family lives separately, all the farm and domestic work is carried out as a communal activity and all property is owned communally. The mean size of their communes is a little over 100. This is because they always split them when they reach a size of 150. The reason they do so is particularly significant from our point of view. The Hutterites say that once a community exceeds 150 people, it becomes increasingly difficult to control its members by peer pressure alone. With smaller groups, a quiet word in a corner of a field is enough to persuade an offender not to behave badly in future. But with larger groups, that quiet word is more likely to elicit a brusque and dismissive response.

Indeed, there is a well-established principle in sociology suggesting that social groupings larger than 150–200 become increasingly hierarchical in structure. Small social groups tend to lack structure of any kind, relying instead on personal contacts to oil the wheels of social intercourse. But with more people to coordinate, hierarchical structures are required. There must be chiefs to direct, and a police force to ensure that social rules are adhered to. And this turns out to be an unwritten rule in modern business organization too. Businesses with fewer than 150–200 people can be organized on entirely informal lines, relying on personal contacts between employees to ensure the proper exchange of information. But larger businesses require formal management structures to channel contacts and ensure that each employee knows what he or she is responsible for and whom they should report to.

Once the significance of this finding had sunk in, I began to find other examples all over the place. And so did other people. Here's one that the American primatologist John Fleagle discovered one day while idly browsing through the Mormon Museum in Salt

Lake City. When Brigham Young was preparing to lead the
Mormons out of Illinois on the Great Trek that culminated in the
founding of Salt Lake City and the Mormon state of Utah, he realized he had a problem. Trying to co-ordinate the activities of some
5000 souls would be a near impossible task. So he divided them
into smaller groups that could operate independently of each other and co-ordinate the activities of their members with maximum
efficiency. He chose groups of 150 as the ideal size.

Sociologists have long recognized that individuals have a limited network of acquaintances. Even in a modest-sized town, an
individual will know only a tiny proportion of those around him
by name or face; he will know even fewer of these well enough to
consider them genuine members of his social circle. Attempting to
estimate the size of this circle of friends is not easy. However, one
quite successful way of doing so involves what are known as
'small world' experiments. The name derives from the discovery
that sending a message to any random individual anywhere in the
world through a chain of personal contacts typically requires only
six intervening steps. If 150 people know 150 other people, then
six steps would allow you to reach 150^6, which is approximately
10 million million people. That's quite a bit more than the 5000
million or so currently alive today. Of course, most people's circles of friends overlap, so that the total number of people who can
be reached in six steps is very much less. Even so, we can reach
5000 million in six steps if each person's network of 150 included
as few as 32 new people who were not in the network of the
previous person in the chain.

During the 1980s attempts were made to estimate people's networks of acquaintances using this approach. The procedure is as
follows: you ask a subject to send a message through a chain of
hand-to-hand contacts to a fictional but realistic-sounding person
somewhere else in the world. The idea is to start by giving the
message to a relative or friend who might, by virtue of their job or
contacts, be able to act as the first link of such a chain. A typical
task might involve sending a letter to a 32-year-old bank clerk
named Juanita living in Mexico City, or a 55-year-old hotel night-
porter named Jim in Sydney. The subject searches through his or

her list of acquaintances for people whose jobs might be relevant. There is Uncle Edward who is an airline pilot, who might know someone else in his company flying to Mexico City, where the letter could be passed on to someone who might have a contact with the right bank. Or they have a friend named Susan who works for a multinational company with mining interests in Australia; Susan might be able to pass the letter on to a colleague travelling to Sydney on company business, where local contacts would ensure that the letter reached the right hands.

Long-suffering subjects are given hundreds of these tasks one after the other. Initially, each new task produces a new name, but after a while the number of new names being added to the list starts to decline, and eventually grinds to a halt. At this point, you have exhausted the subject's circle of friends and acquaintances. Notice that this definition of friends and acquaintances is particularly interesting from our point of view: it represents the circle of people whom you feel you know well enough to be able to ask a favour of them. Two 'small world' experiments of this kind carried out in different US cities yielded estimates that averaged around 135 people. That's encouragingly close to our predicted value of 150.

Church congregations are another interesting example. A recent study commissioned by the Church of England concluded that the ideal size for congregations was 200 or less. This provided a compromise between a group large enough to support the activities of a church, yet small enough for everyone to know everyone else sufficiently well to form a close-knit, mutually supportive community.

The military turns out to be an especially instructive source of information about human group sizes. The army, as everyone knows, is a hierarchically organized system of command in which individual soldiers are incorporated into progressively more inclusive units: first platoons, then companies, then battalions, regiments, divisions, brigades and so on. Military units are interesting precisely because they are subject to intense selection pressure; on the battlefield, men's lives depend on the effectiveness with which the units can co-ordinate their activities. A great deal depends on trust – that the next man will do his bit in the grand scheme.

The smallest military unit that can stand alone is the company. Originally an independent body of men raised by a particular individual, the company developed out of the wars of the sixteenth century as the basic building block of military organisation. Often hiring itself out as a mercenary group, it was as much a way of life as a fighting unit, a loose conglomeration of like-minded men – a bunch of guys who spent time together and, to some extent at least, enjoyed each other's company.

At this early stage, military companies were of varying sizes and effectiveness, depending largely on the size of the leader's purse. During the Thirty Years' War in the seventeenth century, the structure of conventional armies underwent an important reorganization at the hands of the Swedish king Gustavus Adolphus. He established the company as the basic fighting unit with a more or less fixed size; initially 106 men, which is just at the bottom end of the statistical margin of error around the estimate of 150 given to us by the primate brain/group size equation. With the passing centuries, the company consolidated its position as the fighting backbone of the army, but its size grew steadily as new weapons (such as heavy machine guns) and new functions (headquarters unit, medical units) were added. By the end of the nineteenth century, most modern armies were organised along similar lines. During the Second World War, the company had more or less stabilized in size at around 170 men (ranging from Britain at the low end with 130 men to the USA at the high end with 223 men). These values nestle comfortably around the predicted group size of 150, being within the permitted margin of error.

This suggests that, over the years, military planners have arrived at a rule of thumb which dictates that functional fighting units cannot be substantially larger than 200 men. This, I suspect, is not simply a matter of how the generals in the rear exercise control and co-ordination, because companies have remained obdurately stuck at this size despite all the advances in communications technology since the First World War. Rather, it is as though the planners have discovered, by trial and error over the centuries, that it is hard to get more than this number of men sufficiently familiar with each other so that they can work together as a functional unit.

At this size, orders can be implemented and unruly behaviour controlled on the basis of personal loyalties and direct man-to-man contacts. With larger groups, this becomes impossible. Loyalties are no longer man to man, but have to be displaced to more nebulous and less inspiring concepts such as 'the regiment' or 'the Queen'; orders are no longer taken on trust from known individuals, but must be issued by formally created 'ranks' that establish an individual's entitlement to respect. These orders have to be formally signed by an appropriate authority – in former times, they even had to have that person's seal impressed into wax. At the company level, word of mouth continues to be sufficient because everyone knows who everyone is – or at the very least, they know someone who can vouch for every one else.

It is, of course, very easy to play the numerologist and find numbers to fit whatever size your theory requires. Nonetheless, the extent to which the values for these kinds of groups coalesce around 150 is impressive, particularly since all the examples I have discussed share similar bases in terms of social dynamics. Moreover, this value is quite distinct from other well-known human group sizes that are thought to have some cognitive or psychological basis.

One such is the so-called sympathy group size: the number of people with whom you can simultaneously have a deeply empathic relationship. Studies which ask people to list the names of everyone whose death tomorrow they would find devastating consistently yield totals of 11 to 12. Similarly, studies which ask people to list the names of their intimates – say, those friends and relations whom they contact at least once a month – typically yield values in the order of 10 to 15 (these include two of our own studies in the UK.) It is striking that groups of this size are common in situations where very close co-ordination of behaviour is required: juries, the inner cabinets of many governments, the number of apostles, the size of most sports teams.

These sympathy groups are clearly not the same as the 'neocortex' groups of 150 we discussed earlier. It is equally clear that neither is the same as the number of people whose faces you can put names to. There is evidence to suggest that the upper limit to this is around 1500–2000 – well above the neocortex group size, and suspiciously close to the size typical of many tribal groups in traditional societies. The constraint in this case is clearly one of memory capacity, whereas sympathy and neocortex groups are limited by the way in which you relate emotionally to people. Taken together, these results suggest that human societies contain buried within them a natural grouping of around 150 people. These groups do not have a specific function: in one society they may be used for one purpose, in another society for a different purpose. Rather, they are a consequence of the fact that the human brain cannot sustain more than a certain number of relationships of a given strength at any one time. The figure of 150 seems to represent the maximum number of individuals with whom we can have a genuinely social relationship, the kind of relationship that goes with knowing who they are and how they relate to us. Putting it another way, it's the number of people you would not feel embarrassed about joining uninvited for a drink if you happened to bump into them in a bar.

Thus it seems that, even in large-scale societies, the extent of our social networks is not much greater than that typical of the hunter-gatherer's world. We may live in the centre of enormous modern conurbations like New York or Karachi, but we still know only about the same number of people as our long-distant ancestors did when they roamed the plains of the American Midwest or the savannahs of eastern Africa. Psychologically speaking, we are Pleistocene hunter-gatherers locked into a twentieth-century political economy.

All this raises an interesting puzzle. Grooming seems to be the main mechanism for bonding primate groups together. We cannot be sure exactly how it works, but we do know that its frequency increases roughly in proportion to the size of the group: bigger groups seem to require individuals to spend more time servicing their relationships.

If this is so, then we have a problem. The largest typical group size (that is, the average for a species) is the 50–55 characteristic of baboons and chimpanzees, and they seem to be pushing at the limits of the amount of time that can be devoted to grooming without digging disastrously into ecologically more important components of the time budget (such as feeding and travelling time). If modern humans tried to use grooming as the sole means of reinforcing their social bonds, as other primates do, then the equation for monkeys and apes suggests we would have to devote around 40 per cent of our day to mutual mauling. Quite a thought – an almost continuous opiate high.

But no species that has to earn its living in the real world (as opposed to nipping down to the corner supermarket for the week's shopping) could possibly sustain such a heavy investment of time in grooming. It would starve in the process. And this raises an interesting thought about the way we establish and service our relationships. Our ancestors must have faced a terrible dilemma: on the one hand there was the relentless ecological pressure to increase group size, while on the other time-budgeting placed a severe upper limit on the size of groups they could maintain. It seems that somehow they managed to square the circle.

The obvious way, of course, is by using language. We do seem to use language in establishing and servicing our relationships. Could it be that language evolved as a kind of vocal grooming to allow us to bond larger groups than was possible using the conventional primate mechanism of physical grooming?

Language does have two key features that would allow it to function in this way. One is that we can talk to several people at the same time, thereby increasing the rate at which we interact with them. If conversation serves the same function as grooming, then modern humans can at least 'groom' with several others simultaneously. A second is that language allows us to exchange information over a wider network of individuals than is possible for monkeys and apes. If the main function of grooming for monkeys and apes is to build up trust and personal knowledge of allies, then language has an added advantage. It allows you to say a great deal about yourself, your likes and dislikes, the kind of person you are; it also allows you to convey in numerous subtle ways something about your reliability as an ally or friend.

Bonding is a tricky business, because you are committing yourself to a relationship with no guarantee that your partner will reciprocate. You are vulnerable to being cheated by free-riders, who exploit your good nature and then abandon you just at the moment when you most need their help. Being able to assess the reliability of a prospective ally becomes all-important in the eternal battle of wits. Subtle clues provided by what you say about yourself – perhaps even how you say it – may be very important in enabling individuals to assess your desirability as a friend. We get to know the sort of people who say certain kinds of things, recognizing them as the sort of people we warm to – or run a mile from.

Language has an additional benefit invaluable in these circumstances. It allows us to exchange information about other people, so short-circuiting the laborious process of finding out how they behave. For monkeys and apes, all this has to be done by direct observation. I may never know that you are unreliable until I see you in action with an ally, and that opportunity is likely to occur only rarely. But a mutual acquaintance may be able to report on his or her experiences of you, and so warn me against you – especially if they share a common interest with me. Friends and relations will not want to see their allies being exploited by other individuals, since a cost borne by an ally is ultimately a cost borne by them too. If I die helping out a scoundrel, my friends and relations lose an ally, as well as everything they have invested in me over the years. Language thus seems ideally suited in various ways to being a cheap and ultra-efficient form of grooming.

The conventional view is that language evolved to enable males to do things like co-ordinate hunts more effectively. This is the 'there's a herd of bison down by the lake' view of language. An alternative view might be that language evolved to enable the exchange of highfalutin stories about the supernatural or the tribe's origins. The hypothesis I am proposing is diametrically opposed to ideas like these, which formally or informally have dominated everyone's thinking in disciplines from anthropology to linguistics and palaeontology. In a nutshell, I am suggesting that language evolved to allow us to gossip.

CHAPTER 5

The Ghost in the Machine

One of the more curious uses to which we put language is poetry and song. I shall return to song in more detail later, but poetry provides us with an unexpected window on how language works. Consider the following poem:

> A *hot and torrid bloom which*
> *Fans wise flames and begs to be*
> *Redeemed by forces black and strong*
> *Will now oppose my naked will*
> *And force me into regions of despair.*

The remarkable thing about this poem is that it was composed by a computer program. The program, known by the acronym RAC-TER, was written by a New York computer buff named Bill Chamberlain, and the series of poems and short essays it produced was published in 1984 as a small volume entitled *The Policeman's Beard is Half-Constructed.*

What makes this book particularly relevant to my theory is the fact that the computer obviously had no idea what it was doing. Indeed, it was not trying to do anything specific, in the sense that a human might sit down and try to write a poem. The program was simply designed to find words that fitted together grammatically. It consisted of a dictionary of English words which specified whether each word was a noun, verb, adjective, adverb and so on. The program looked up a word at random in its dictionary, then checked to see if it fitted grammatically with the word(s) immediately before it. If it didn't, the program rejected the word and tried another one; if the word fitted grammatically, it added the word to the sentence and moved on to the next space.

Remarkably, we can read these writings and make perfectly

coherent sense of them. To be fair, the essays are at times a little strained – although an ounce of generosity on the reader's part should enable them to see something meaningful even in these. But the poems are really quite passable. Indeed, the book was given surprisingly good reviews in the highbrow newspapers.

What this highlights, I think, is just how much of language is in fact *communication* – someone actively trying to influence the mind of another individual. The human mind seems to have been built in such a way that it assumes other individuals are trying to communicate with it. We interpret their body language, their signals and their speech not so much by what the words themselves say in a literal sense, but in terms of what we suppose lies behind them. 'Just *what* is he trying to tell me,' you ask as you struggle to make sense of someone's barely coherent (but grammatically correct) speech. In other words, we assume that everyone else behaves with conscious purpose, and we spend much of our time trying to think our way into their minds so as to divine their intentions. We are so imbued with this way of viewing the world that we easily transpose it on to other animals, and even occasionally on to the inanimate world.

Natural though this way of thinking may be for the man in the street, philosophers and scientists have tended to make rather heavier weather of it. In particular, they have expressed grave misgivings about the concept of consciousness, and have often insisted that it doesn't exist outside humankind. The history of it all begins with the hugely influential seventeenth-century French philosopher and mathematician René Descartes.

Descartes' Dilemma

Descartes was much taken by the idea that mechanical models could be built that opened doors and played musical instruments, and he designed and built several himself. The models held an important philosophical lesson for him. If we could build models that acted so realistically, we should be cautious about attributing a mental life to animals whose behaviour impressed us with its thoughtfulness. We know that humans probably have an inner mental life because they speak, and what they say resonates with what we ourselves experience. But animals do not speak, and the sensible conclusion is that they do not have minds (or souls) in quite the same sense as we do. They certainly feel and have emotions, but these could simply be mechanical responses to the stimuli that impinge on them.

Descartes bequeathed to us a clear view of the difference between humans and other animals. We have minds; clever and beguiling as they may be, animals are mere machines. That judgment has coloured not only how we view animals, but also how we treat animals – not to mention their legal status and their rights under the law. We can experiment on animals in a way that we are not allowed to do on humans. The Cartesian view, as it is often called, has underpinned some three hundred years of medical science.

The second half of the nineteenth century saw a considerable revival of interest in the behaviour of animals. Darwin's great revolution in biology had placed humans on the same continuum as animals. This inevitably led to Darwin himself and many of his contemporaries looking anew at the emotional and mental lives of animals. There were so many obvious similarities between their behaviour and our own that a common evolutionary origin for the emotions seemed an obvious conclusion.

Unfortunately, Darwin and his contemporaries were somewhat limited in their sources of information about the behaviour and mental states of animals. Most of what they had to go on were the observations of natural historians and their own casual experiences. Their books are littered with remarks such as, 'Colonel So-and-so informs me that his favourite fox-hound once ...' As the claims based on such evidence became increasingly extravagant, a reaction inevitably set in. By the early years of this century, a consensus emerged which placed a very firm lid on the idea that animals (and perhaps even humans) had minds. We cannot see minds, the argument ran, but we can see behaviour; so rather than speculate about unverifiable mental events, a science of the mind should deal only with observable behaviour whose existence could be independently confirmed. And so was born the school of psychology known as 'Behaviourism'. It was to dominate the psychological sciences from the turn of the century until the 1980s.

Behaviourism served a very valuable purpose in the developing science of the mind because it forced everyone to be more rigorous in their evaluation of the phenomena they wanted to discuss. If nothing else, it certainly curbed the excesses of late-Victorian fantasy. But with the benefit of nearly a century of intensive experimental studies in psychologists' laboratories and half a century of careful ethological study of animals in the wild, we are now in a position to re-evaluate Descartes' claim that animals are simply machines.

During the last fifteen years there has been a radical rethink of the evidence for the mental states of animals. We have learned how to ask our questions better, and we have learned how to tease answers out of those who do not have language, be they human or animal. The jury's final verdict has yet to be handed down, but there is enough evidence to suggest that the story is more complex and more interesting than Descartes could ever have imagined.

The deliberations of the last decade have shown that the central issue is something that psychologists now rather confusingly refer to as a 'Theory of Mind' (or ToM for short). Having a Theory of Mind means being able to understand what another individual is thinking, to ascribe beliefs, desires, fears and hopes to someone else, and to believe that they really do experience these feelings as mental states. We can conceive of a kind of natural hierarchy: you can have a mental state (a belief about something) and I can have a mental state about your mental state (a belief about a belief). If your mental state is a belief about my mental state, then we can say that 'I believe that you believe that I believe something to be the case'. These are now usually referred to as orders of 'intensionality'.[1] Thinking about mental states in this way yields the following rough hierarchy.

1. Intensionality in this sense is usually (but not always) spelt with an 's' to distinguish it from conventional intentions (with a 't'), which are simply one kind of 'intension'. Although some people have recently dropped the distinction, I prefer to retain it since it avoids unnecessary confusion.

Machines such as computers have zero-order intensionality: they are not aware of their own mental states. Presumably, we also have zero-order intensionality when we are in a coma, and most insects and other invertebrates are also probably zero-order intensional beings. Ever since Descartes produced his immortal aphorism *Cogito ergo sum* ('I think, therefore I am'), we have known about first-order intensional states (I believe something to be the case). After that, we run into the beginning of an infinite regression: I believe that you believe something (second order intensionality), I believe that you believe that I believe something (third order intensionality), I believe that you believe that I believe that you believe something (fourth order intensionality), and so on. For obvious reasons, the higher orders of intensionality are often referred to as 'mind-reading'.

There is good reason to believe that humans are capable of keeping track of, at most, six orders of intensionality, and after that they probably have to see it written down. In the words of the philosopher Dan Dennett: 'I suspect [1] that you wonder [2] whether I realize [3] how hard it is for you to be sure that you understand [4] whether I mean [5] to be saying that you can recognize [6] that I can believe [7] you to want [8] me to explain that most of us can keep track of only about five or six orders [of intensionality].'

This hopelessly convoluted sentence makes the tortuous prose of the typical Victorian novelist seem the very paragon of clarity. In fact, it contains eight orders of intensionality, which I have numbered in brackets. Without seeing the sentence on paper, it would be impossible to work out what is going on, and even on paper we have to cut through some of the complexities to make it comprehensible. I'd bet a pound or two that not many people could reconstruct all the mental states involved, or remember by the end who was doing the suspecting at the start.

My confidence in this matter is not just born of instinct. My colleagues Peter Kinderman, Richard Bentall and I have tested people on these kinds of problem, using stories with up to five orders of intensionality. At the same time, we have tested them with similar stories whose content is a simple sequence of up to six causally related factual events. People typically make a small but steady number of errors with pure event-memory questions, and the frequency remains constant no matter how long and involved the story becomes in terms of the events involved – at least up to six links in the causal chain. Although with ToM stories they behave similarly up to about the third order of intensionality, the number of errors they make starts to rise exponentially from that point on. At fifth order intensionality, the number of errors is more than five times that for comparable event-memory questions.

These results show just how difficult theory of mind tasks are in practice. It is not surprising, then, to find that not all humans are capable of performing at such high levels. During the early 1980s psychologists began to suspect that children were not born with a theory of mind; rather, they acquire one as they develop.

It turns out that children reach a critical watershed at about four to four-and-a-half years of age. They suddenly seem to realize that other individuals can hold beliefs different from their own. Up to that point, children tend to interpret the world (and other people's beliefs about the world) rather as they see it. They cannot imagine what it would be like to believe that the world is other than it seems to them, so they do not realize that you might have different opinions and beliefs from the ones they hold. They assume that you see everything they see, and interpret it in much the same way.

This has important consequences: up to the age of about three, children cannot lie (or at least they cannot lie convincingly). That is to say, they do not seem to be aware that your state of mind, your beliefs, can be manipulated. By the age of about three, they know enough to be aware that if they deny having eaten the chocolates vigorously enough, you will often believe them. But a child of this age does not know enough to be aware that the chocolate smeared around his mouth gives the game away. This is very different from children's behaviour just a few months later, when they have acquired a theory of mind; then they can manipulate you something rotten.

Psychologists have developed a crucial test for theory of mind, known as the 'false belief test'. It asks the key question: is the child aware that someone else can hold a false belief (or at least a belief that the child supposes to be false)? The now-classic example of this is the so-called 'Sally and Ann' test. Sally and Ann are two dolls who are presented to the child and formally introduced. The child is shown that Sally has some sweets, which she then places under a cushion on a chair. Having done this (perhaps with the child's help), Sally leaves the room. Then, while Sally is out of the room, Ann takes the sweets from under the pillow and puts them in the pocket of her dress. When Sally comes back into the room, the child is asked, 'Where does Sally think the sweets are?' Up to the age of four, children invariably answer, 'In Ann's pocket.' But after about four-and-a-half, they invariably say, 'Under the pillow,' adding with conspiratorial glee, 'but they aren't there!'

In another classic test, a child is shown the cardboard tube in which the sweets known in Britain as Smarties and in the USA as M & Ms are sold. When the child is asked what he or she thinks is inside, the answer is of course 'Smarties'. The top is then removed, and the child is shown that inside the tube are some pencils rather than the sweets he or she expected. Finally, the child is told that its best friend is just about to be brought into the room and shown the same tube; what does the child think its friend will say when asked what is in the tube? Up to the age of four, children invariably say, 'Pencils'. But after four or so, they are aware that others can hold beliefs different from their own, and so they will say, 'Smarties'.

Although the appearance of ToM strikes us as a very sudden process in children, in fact it is the outcome of a long process of intellectual experimentation. From a very early age, children become aware that other objects in the world are capable of doing things for you. You can ask them to give you things you want; sometimes you can even blackmail them into doing so by whining and grizzling until you eventually break down their resilience. Experience with the different kinds of objects they encounter leads children to conclude that some of these objects are animate and others inanimate. At first they fail to distinguish clearly between a person and a doll, apparently believing that dolls share with humans all the qualities of volition they see in humans. But with experience, they come to separate these categories out.

By about the age of three, children are into what has been termed 'belief/desire psychology'. They can recognize that other individuals have wants and desires that are similar to those they experience. Over the course of the following year, they use this knowledge to build up a picture of how other individuals tick. It's a very complex task for children, and of course they make many mistakes. It is now clear that understanding the social world is a far more difficult task for children to master than understanding the physical world.

This realization has turned the influential psychologist Jean Piaget's theories of development on their head. Piaget's theories have dominated our thinking on child intellectual development for the better part of half a century. He was much exercised by the desire to explain children's growing understanding of their world. Like all his contemporaries, Piaget assumed that our brains are there to process information about the world, and hence that coming to an understanding of the underlying features of that world was the most difficult task a child has to achieve. The rest, he supposed, was plain sailing.

Piaget, it seems, was deceived by the extraordinary skills of children in the social domain into thinking that these skills were not problems of any great consequence. Children obviously acquired them rather effortlessly early on. Piaget's mistake was understandable: few at that time appreciated just how complex our social world actually is. Social skills are a matter of great urgency for a child: its very survival depends on them. So long as the minor matters of seeing and hearing develop early on, the more complex tasks like understanding the conservation of quantities and volumes – the problems Piaget made so much of – can wait until after the child has managed to negotiate its way through the social maze into which it has been born.

Is There Anybody Else Out There?

Of course, Piaget was not all wrong. He noted, for example, that children are initially self-centred and only gradually come to break the mould of their egocentric universe to take on another individual's perspective. Although the language he used is different, Piaget's understanding of at least this part of the story seems to have been basically right. We are born without theory of mind; with time, we gradually acquire ToM and are able to understand how others feel and think, and so to use this knowledge in our social interactions. So if we set the standard for human-ness by what we see in adult humans, it follows that, strictly speaking, human babies are not fully human, and do not become so until the age of about four.

Of more interest still is the fact that some people never develop ToM. We now refer to these individuals as autistic, a medical syndrome first identified as recently as the 1940s, although it has surely been around for very much longer than that. Autistic people – the great majority of whom are males, indicating a clear genetic component to the condition – vary in the degree of their impairment. Some are very severely handicapped, never develop language and show no ability to interact socially with others. Others develop language, but remain social isolates. Sufferers from its mild form, known as Asperger's Syndrome, can seem quite normal aside from their social gaffes and occasionally bizarre behaviour. Only the most subtle psychological tests allow us to recognize that these individuals may not have full theory of mind.

Autistic people are characterized by two key deficits. One is a consistent failure to pass false belief tests. The other is an apparent inability to engage in pretend play. The psychologist Allan Leslie has argued that these two characteristics are intimately related to each other. Because they do not realize that other people can hold beliefs that are false (or at least beliefs that they *suppose* to be false), they are unable to imagine other worlds or that the world could be other than as it now is. Consequently, they cannot engage in fictive play. They will not, for example, run through the motions of a dollies' tea party; dollies are not living organisms, so how could they possibly do things that real people do? Nor will they pretend to be asleep in order to play a joke on someone else. Nor will they tell a deliberate lie, because lying requires you to be aware that the other person might not know all you know. An autistic person simply assumes that the world is transparent, that he and his audience share the same information. In effect, autistic people take the world exactly as it comes. One consequence is that they fail to recognize the richness of meaning often buried in our use of language. Here is a classic example reported by the mother of an autistic teenager. Before leaving the house to visit a neighbour across the road, she told her autistic son that if he wanted to come on over later, he was to be sure to pull the door behind him. An hour or so later he did exactly that, having first wrenched the front door off its hinges.

This story should remind us of just how much of the meaning in our conversations depends on the listener reconstructing the speaker's mind-state. Autistic people simply cannot do that, because they do not realize that what someone else has in mind may in fact be different from the normal meaning of the words they are using. In fact, almost all of our conversational exchanges are metaphorical or require interpretation by the hearer. We commonly speak in telegraphic fashion, providing just the key points and assuming that the listener can fill in the bits and pieces to make sense of what we say. Here is a classic example of the deep layers of interpretation with which our conversations are often imbued:

HIM: I'm leaving you!
HER: Who is she?

Clearly, his statement could have a dozen different and equally legitimate interpretations that hinge on the context and the past history of the individuals concerned. Yet even with so brief a script, we have no difficulty in identifying instantly the correct interpretation from her elliptical reply. We can at once fill in all the details and provide a great deal of probable background information.

Autistic people could not do that. Those suffering from the milder Asperger's condition often manage to cope with social situations, so passing false belief tests. But they do not do so by mind-reading in the way that we do. They cope by what the psychologist Francesca Happé refers to as 'hacking it'. In other words, they are smart enough – being of normal intelligence – to work out rules of thumb that allow them to make the right decision in a social situation nine times out ten. But they appear to have no idea why these rules of thumb work, merely that they do.

The following musical analogy will perhaps give you a feel for their problem. Much though I like music, I happen to be all but tone-deaf. I can recognize Mozart's *Serenade in E flat* when I hear it, though I couldn't in all honesty tell you whether it's in B flat, A minor or for that matter G sharp. But a musician will know at once which key it's in, even though he's hearing it for the first time. I've learned to recognize the tune as a specific series of notes, but I have no real idea why the key is called E flat and it probably wouldn't matter if I did: even if I tried, I couldn't generalize the rules of recognition I've learned to other pieces. So it is with Asperger's.

I remember the mother of an Asperger's child observing that her son, then about twelve years old, knew that people had friends and that he did not, yet he had no idea how you went about obtaining a friend. Did you buy them at shops, or did you simply say to someone – anyone – 'You are my friend!' It is heartbreaking to encounter cases like this, because there is no way to get through to them the deep emotional bases of normal human relationships. They simply cannot comprehend what they are or how they work. Indeed, they would find my use of the word 'observe' in the first sentence of this paragraph very puzzling: the mother wasn't *seeing* anything at all! Yet it is worth reminding ourselves that these individuals are otherwise completely normal, and may even be of above-average intelligence. Asperger's people are often very good at mathematics, for example, probably because they can think clearly in the abstract and not become confused by emotional and other irrelevant associations.

The obvious question at this juncture is how other animal species compare with us on the intensionality scale. Are we alone in the universe of the mind or is there somebody else out there? The most likely place to find other species that have the same theory of mind abilities as we do is among our closest relatives, the apes.

The realization that theory of mind is in some sense a key to understanding the human mind inevitably provoked considerable interest in how monkeys and apes compare on the intensionality scale. The problem lies in identifying a critical test we can use as an infallible criterion. The earliest attempt at this problem was by the American psychologist Gordon Gallup. He argued that the key factor which characterizes us is that we can recognize ourselves as a separate entity from the other individuals with whom we live (in effect, Descartes' *I think, therefore I am*). Being self-aware provides us with the capacity to reflect on our internal mental states, and then to relate these internal states to the observable behaviour of other individuals. From this comparison, we can work out that other individuals also have mental worlds.

Gallup designed an ingenious test which he claimed allowed us to decide quite unequivocally whether or not another animal is self-aware. The test involves training an animal to use mirrors. Later, the animal is anaesthetized and a small dot of dye put on an area of bare skin on its face. The key question is: once it has recovered from the anaesthetic, will the animal realize there is something different about its face? Can it demonstrate this by touching or picking at the spot of dye?

Over the past decade, extensive experiments of this kind have been carried out on chimpanzees, gorillas, orang-utans and several species of Old World monkeys, as well as on porpoises and elephants. Despite a few contradictory outcomes, the consensus on these results is fairly uncontroversial. Chimpanzees readily solve mirror problems of this kind; orang-utans and gorillas seem to be reasonably competent (though many fewer of them have been tested); but no monkey has yet passed a mirror test. The obvious conclusion everyone has reached is that great apes are self-aware, but that other members of the Primate Order (and that appears to include the gibbons, the so-called lesser apes) are not. When similar experiments were tried with elephants, the animals apparently failed even to recognize the wall-size mirrors as mirrors, but tried to walk through what they thought was an open door.

Although many have been tempted to conclude that the great apes have theory of mind and monkeys do not, some lingering doubts about Gallup's experimental design niggle at the back of one's mind. Why on earth should the ability to use a mirror be a convincing test of self-awareness? After all, monkeys and apes don't have mirrors in the wild, so why should the ability to recognize oneself in a mirror demonstrate the ability to appreciate that you have an independent mental life? Just what, in fact, do the results of the mirror tests tell us?

One obvious answer is that they merely tell us whether a species is smart enough to understand the physics of mirrors. It may be, for example, that technical problem-solving skills are in some way spun off social intelligence: one way this might come about is if advanced mentalizing skills require a lot of computing power, and this computing power allows you to solve the rather trivial physical problems associated with mirrors.

Gallup's mirror test is certainly telling us something, but it just isn't clear what. We need something that is more diagnostic of mentalizing abilities. One idea proposed by the psychologists Dick Byrne and Andrew Whiten is tactical deception. This is the name given to occasions when one individual tries to exploit another by manipulating its knowledge of the situation. One example would be the juvenile baboon Paul manipulating its mother in order to take Mel's tuber; another would be Hans Kummer's young female hamadryas baboon inching her way to sit grooming with the young male of her choice behind a rock without arousing her harem male's suspicions (see page 23).

Kummer noted another form of deception when studying gelada baboons in captivity. Gelada also form tightly bonded harems, and male harem-holders are as unhappy as hamadryas males about their females straying too far from them. One day, the harem male was removed from the group and put in a cage out of sight of the compound where the group lived, but not out of auditory contact; the harem male could still hear everything that happened in the compound, and the rest of the group could still hear and interact vocally with him. With the old male out of the way, the group's young follower male made the most of his opportunities and began to mate with one of the females. Kummer noticed that when they did so, the male and the female both suppressed the raucous calls that gelada normally give at the climax of mating – calls that can normally be heard at distances of 100 yards or more. Kummer referred to this as 'acoustic hiding'.

Similar behaviour has been reported from captive chimpanzees. Frans de Waal once watched a female who was mating surreptitiously behind some bushes with a low-ranking male place her hand over the male's mouth to prevent him giving the loud copulation calls that would have been heard by the dominant males on the other side of their large compound.

In both cases, the offending couples seemed to be trying not to give the game away by vocalisations that would be clearly audible over the fence. This is tactical deception: the animals are apparently trying to suppress clues about what is happening, so that other animals remain in ignorance.

Another form of tactical deception was described by Sue Savage-Rumbaugh. She has been engaged in a long study of the two chimpanzees, Austin and Sherman, who were taught an artificial keyboard language. Sherman was inclined to bully Austin, much to Austin's distress. One day, Austin discovered that Sherman was afraid of noises from outside their sleeping quarters, especially at night. Thereafter, whenever Sherman's bullying got too much to bear, Austin would race into the outdoor part of their accommodation, bang vigorously on doors and other objects, then rush back in whimpering and doing his best to look terrified. Sherman invariably responded with panic and would ask to cuddle Austin for comfort.

Tactical deception provides something of a bench mark for advanced mentalizing abilities because it requires at least second-order intensionality. In order to engage in tactical deception, an animal has to be capable of appreciating that its opponent believes something to be the case. By altering the information available to the opponent, the deceiver attempts to influence the beliefs of the opponent: I have to understand that, by behaving in a certain way, you will believe that I am doing something innocuous. And that obviously involves holding a false belief of the kind we discussed earlier.

Whiten and Byrne collected a large database of examples of tactical deception from the primate literature and from odd instances observed by colleagues while studying various species. Their most interesting finding for us is that instances of tactical deception are virtually absent from the Prosimians (lemurs, galagos, etc.) and rare among New World monkeys. They are common among the socially advanced Old World monkeys (baboons, macaques), but most of the instances reported to them come from chimpanzees (with a handful from the other, much less intensely studied great apes).

In fact, Dick Byrne later compared an index of the frequency with which tactical deception was reported among the species on their database with my index of relative neocortex size and found a very good fit. Species with large neocortices and complex societies, such as chimpanzees and baboons, were much more likely to engage in tactical deception than species such as African colobus monkeys or South American howler monkeys, which had much smaller neocortices and rather small social groups. It seems that a minimum of computing power is necessary to think through the complications involved in tactical deception.

Following the logic of Byrne's analysis, it occurred to my Polish colleague Boguslaw Pawlowski and me that if the Machiavellian Intelligence hypothesis really did work, we should be able to show similar correlations between neocortex size and features like the stability of the male dominance hierarchy. We reasoned that, in species with large neocortices, low-ranking males would be able to take advantage of more subtle social strategies, such as tactical deception, to circumvent the dominance of the higher-ranking males. In the normal course of events, high-ranking males are able to monopolize females during the mating season, so preventing lower-ranking males from mating. The result is commonly a straightforward relationship between a male's rank in the hierarchy and the frequency with which he is able to mate or sire offspring.

Pawlowski and I reasoned that, if low-ranking males with big brains could exploit loopholes in the system, we should find that the relationship between male rank and reproductive success becomes less strict as neocortex size increases. This is exactly what happened. Since one of the strategies males are using in these contexts resembles tactical deception, our results reinforce Byrne's findings and confirm that the use of tactical deception actually does have functional consequences, in this case by affecting the reproductive success of males and so ultimately their genetic fitness (their contribution to future generations).

Byrne's findings on tactical deception and ours on male mating strategies provide strong behavioural evidence to support the Machiavellian Intelligence hypothesis. It suggests that the ability to use subtle social strategies and to exploit loopholes in the social context depends on how much computing power you have available in your brain.

These findings do not tell us, however, how the differences we see between species relate to differences in levels of intensionality. All we know is that chimps can do better than baboons, and baboons can do better than howler monkeys. But where do they all stand in terms of their ability to think reflexively about the contents of other individuals' minds? Is a baboon better than a howler monkey because it can imagine what it is like to be another baboon, or simply because it can carry out more complex calculations about the consequences of a social act?

At this point, we reach the limits of our current state of knowledge. No one has so far been able to relate levels of intensionality to particular behaviour patterns in any detail. We do, however, have some clues that at least point us in the direction of the likely outcome. They are mostly anecdotal observations, but they are interesting for all that.

Vicki, the chimpanzee who was raised by the Hayes family alongside their own child during the 1950s, was once observed to walk along pulling a piece of string behind her. At first sight, this seems innocent enough. But when the end of the string reached a step between two floor levels, she stopped and exhibited signs of consternation. Her behaviour was exactly what you might expect from a child who was pulling a toy car on a piece of string when it got jammed. She went back to the end of the string and lifted it carefully over the step, as though freeing it from the obstruction. That done, she went on her way again. This has all the hallmarks of fictional play, of the kind Alan Leslie has identified as a key feature associated with theory of mind, and which is absent in autistic children.

Another example of high-level intensionality was observed by the Dutch ethologist Frans Plooij while studying Jane Goodall's chimpanzees in the Gombe National Park in Tanzania. At that time, the researchers were using bananas and other foods to attract the chimps into the camp area so that they could be studied more easily. As time went by, the chimpanzees became progressively more demanding. Worse still, having discovered where the bananas were stored, they began raiding the huts and storehouses at the camp. To prevent things getting completely out of hand, the researchers built a concrete box half-buried in the ground. The box had a lid which the researchers could release from a short distance away by means of a cable. In this way, the researchers hoped to ensure that low-ranking animals gained a fair share of bananas, and wouldn't be discouraged from coming to the camp by the fact that more dominant animals prevented them from getting any of the food on offer.

One day, one of the lower-ranking males arrived alone at the feeding point. The catch was released with an audible click to allow him to open the lid of the box and feed on the bananas inside. Just as he was about to do so, however, one of the dominant males appeared. The first male at once pretended to show no interest in the food box. This was a reasonable ploy: because the catch was only released when a specific individual was present, it often remained locked even when chimpanzees were in the feeding area. Presumably, the male in this particular instance wanted to give the impression that the box was still locked and hence that there was little point in the other male hanging around. Tactical deception of this kind is well within the range of a chimp's abilities: both de Waal and the American psychologist Emil Menzel have observed chimps behaving in just this way in their respective captive colonies in Holland and the USA. But the fascinating point about this incident was the behaviour of the dominant male. Rather than investigating the banana box for himself (which would have been a pointless task anyway), he turned and walked away again; but when he came to the edge of the clearing, he slipped behind a tree and peered back to see whether the male at the feeding box tried to lift the lid after he had gone.

If my interpretation of the behaviour of these chimps is correct, the dominant male was behaving in a way that clearly implies at least third-order intensionality. Something like this must have been going through his mind: I think [1] that Jim is trying to deceive [2] me into believing [3] that the lid is locked. It's just conceivable he might even have been indulging in fourth-order intensional thinking: I think [1] that Jim is trying to deceive [2] me into believing [3] that Jim thinks [4] the lid is locked.

The trouble with anecdotes of this kind is that it is always possible to provide an alternative explanation in terms of coincidences or simpler learned behaviours. Did Austin *really* understand that Sherman was scared of noises in the dark? Or had he simply learned that if he made a lot of noise outside, Sherman would cuddle him instead of bullying him – even though he didn't really know *why* Sherman preferred to cuddle rather than bully in these particular circumstances? Did the male at the food box really *intend* to deceive his rival – in effect saying to himself, 'If I behave in a nonchalant way, I think [1] this male will believe [2] that I think [3] the box is still locked' – or had he merely learned that by behaving in this way, rivals would eventually go away, for some reason completely beyond his powers to fathom? Perhaps the dominant male in this last story stopped as an afterthought because he couldn't quite tear himself away from the food box in case the catch was released – as sooner or later it would eventually be – and happened to be behind a tree when he did so? After all, the chimps may not have worked out why the lid would open on some occasions but not on others; rather, they may have learned that patience was eventually rewarded by food if you returned to the box often enough.

We would feel surer of our ground if we could point to a long series of examples all of which showed the same kind of behaviour. The more examples we had, the less likely it would be that they were all the result of coincidence; although it would still be difficult to exclude simpler explanations in terms of straightforward pattern learning.

It is nonetheless interesting that observations of this kind have come only from chimpanzees. Despite the hundreds of thousands of hours scientists have spent studying Old and New World monkeys in the wild and in captivity, events that can be interpreted at high levels of intensionality have never been reported (although this could, of course, simply mean that observers have not noticed such instances because they have not expected to see them).

In the last few years there have been several studies which attempted to circumvent this difficulty by focusing more clearly on experimental designs that mirror the kinds of tests used to establish whether children have ToM.

The earliest tests were carried out on a language-trained chimp named Sarah by the psychologist David Premack in the early 1980s. Premack and his colleague Guy Woodruff showed Sarah clips of film of someone trying unsuccessfully to do something: for example, reach a banana suspended from the ceiling. They then offered her photographs of appropriate and inappropriate solutions to the problem. An appropriate solution might be a set of boxes piled up one on top of the other below the banana, while an inappropriate solution might be the same boxes lying scattered on the floor. Sarah displayed considerable competence at understanding the person's intentions, as shown by her picking the appropriate solution more often than not. From experiments like these, Premack and Woodruff concluded that Sarah's ability to understand another individual's intentions demonstrated that she had, in some sense at least, a theory of mind.

Two more recent series of tests have tried to compare apes with monkeys to see whether there are any differences between these closely related members of the Primate Order. In the first of these, the American psychologist Danny Povinelli put chimps and rhesus macaques (a representative advanced Old World monkey) through a series of tests designed to discover whether they understand another individual's intentions or knowledge of a situation.

Among the tests he gave chimpanzees, for example, was one in which the animal had to choose between two humans in order to get a reward it could not reach: a glass of juice to drink. The chimp was shown pictures of two assistants, and had to choose between them by pushing over the appropriate holder in which the photographs were held. The difference between the two humans was that one of them always deliberately poured the juice on to the floor, whereas the other one did so accidentally, for example by dropping the cup when picking it up or by tripping when handing it to the chimp. Could the chimp distinguish between deliberate and accidental behaviour? The answer seemed to be a clear yes: the chimp soon learned to choose the human that accidentally spilled its juice.

In another series of experiments, the chimp was given the opportunity to obtain a food reward from a baited box that was out of its reach. In order to get the reward, the chimp had to choose a human assistant to open the box and hand it the food. Two assistants pointed at different boxes, and the chimp had to decide on which assistant it thought was more likely to be right. The choice was between an assistant who was in the room and watched the box being baited, and an assistant who conspicuously left the room while the box was baited. The first assistant knew where the reward was, but the second obviously did not. So the correct answer was to choose the box which the knowledgeable assistant pointed to. Most (but not all) of Povinelli's chimpanzees managed to solve this problem reasonably competently, but none of the monkeys did. It seems that apes, or at least chimps, can distinguish between knowledge and ignorance in other individuals, but monkeys cannot. However, even though they outshone the monkeys, the chimps do not seem to be as competent at these tasks as human children are.

This was borne out by another series of tests, carried out by Sanjida O'Connell, one of my students. She designed a mechanical version of a false belief test which aimed to meet the standards of the Sally-and-Ann test used on children. Once again, the animal was presented with a choice of four boxes. The experimenter placed a peg above one of the boxes; then she placed a morsel of food in the box she had previously marked with a peg. The chimp was then free to open the box of its choice and collect the reward if it chose the right box. Once the chimp had learned to respond correctly to the basic procedure, a glitch was introduced. The boxes were so designed that O'Connell could not see the front of them when she was baiting the selected box from the back.

Having first put the peg in place from the chimpanzee's side of the apparatus, she then went round the back to bait the box, just as she had done dozens of times before. Only this time, while she was on the other side, the peg moved – apparently of its own accord, but really by means of a lever she surreptitiously operated – and came to rest above another box. The critical question was: would the chimp recognize that the box which had been baited was the one the experimenter had originally identified with the peg – presumably the box the experimenter *thought* was the right box – or would it assume that the experimenter had the same knowledge as it did, and so would bait the box above which the peg *now* rested). This is about as close to the Sally-and-Ann test as it is possible to get.

Although the chimps did better than autistic adults at this task, they were nowhere near as good at it as five and six-year-old normal children (who have ToM). The chimpanzees certainly learned to solve the problem, but they weren't as competent as one might have expected had they had full ToM.

There is, however, one final caveat. Negative answers are never very satisfactory when working with animals, especially chimpanzees; you can never be sure whether failure to perform to a criterion really does indicate an inability to solve the problem we set for them, or simply a lack of interest. Not surprisingly perhaps, chimps often seem to find tests of these kinds boring, and sometimes they cannot be persuaded to take part at all. There is a lovely sequence in the BBC Horizon film *Chimp Talk* in which Sue Savage-Rumbaugh asks Kanzi to carry out a series of instructions. When it came to 'Put the bunch of keys in the refrigerator', you can sense in his momentary hesitation the puzzlement that must have been in his mind: 'What *is* she up to now? Oh well, I suppose I had better humour her, poor thing!'

Our tentative conclusion must be: there is sufficient evidence to suggest that chimpanzees, if not all great apes, have ToM in some form – even though it may be a less advanced form than in humans – but that monkeys do not. Even though monkeys are more advanced in these respects than other animals, they certainly do not have full ToM. Rather, they probably have the kind of cognitive levels reached by three and four-year-old children just before they finally develop ToM.

Dorothy Cheney and Robert Seyfarth have commented that monkeys are good ethologists but bad psychologists: they are good at reading another individual's behaviour but bad at reading its mind. They give a delightful illustration of this from their studies of the vervet monkeys in Amboseli, Kenya. One day, a strange male appeared in a grove of trees not far from the troop they were studying. Lone males of this kind are invariably intent on joining a group, and are usually able to displace the group's dominant male when they do so. For the incumbent male this is not a happy occasion, because he stands to lose his monopoly over the females. Naturally, they resist intruders by every means they can. In this case, the incumbent was positively inspired. As the intruder stepped down from his tree to cross the open ground to the trees the group was feeding in, the troop male gave a leopard alarm call. The intruder shot back into the safety of his tree. Later, satisfied that all was well, he tried again; once again, the troop male gave his leopard call. So far, so good: the ruse seemed to be working. Unfortunately, the troop male eventually gave the game away by giving his leopard call while he himself was walking across the open ground. The intruder was smart enough to realize that no one in his right mind gives alarm calls while wandering nonchalantly across open ground where he is at risk of being caught.

Into the Mind and Beyond

Theory of mind is, beyond question, our most important asset. It is a remarkable skill. Yet even ToM pales into insignificance by comparison with where this skill has taken us.

ToM has given us the crucial ability to step back from ourselves and look at the rest of the world with an element of disinterest. The starting point of all this was probably our ability to reflect back on the contents of our own minds. Why do I feel the way I do? Why am I angry now? Why do I feel sadness or happiness? Understanding our own feelings is crucial to understanding those of other people. Without recognizing what we are seeing in others, we have no hope of delving into their minds sufficiently far to appreciate their mental reactions to the things they experience.

The real breakthrough is where fully developed third-order ToM allows us to imagine how someone who does not actually exist might respond to particular situations. In other words, we can begin to create literature, to write stories that go beyond a simple description of events as they occurred to delve more and more deeply into why the hero should behave in the way he does, into the feelings that drive him ever onwards in his quest.

I think I'm on safe ground in arguing that no living species will ever aspire to producing literature as we have it. This is not simply because no other species has a language capacity that would enable it to do this, but because no other species has a sufficiently well-developed theory of mind to be able to explore the mental worlds of others. To write fiction is to conceive of imaginary worlds that do not exist. I am not convinced that even the ToM abilities of chimpanzees are good enough to be able to do that. Chimps seem to be limited at most to third-order intensionality ('I believe that you want me to think that the food box is locked'). Humans seem to be capable of following arguments through to fourth-order intensionality without too much difficulty, though they do not very often run to such lengths in everyday contexts.

But they do become necessary when writing stories whose plots involve both the writer and the reader understanding [1] what one character thinks [2] another character wants [3] the first character to believe [4]. Since both writer and reader become part of the chain of intensionality, they must be able to go one order beyond what the characters actually do. To keep track of that through the sequence of events in a novel is obviously very demanding. The writer has to be able to assume that his readers can achieve the same levels of intensionality as he can; if the reader was incapable of that, there would be little point in trying to sell the novel to a publisher.

The ability to detach oneself from the immediacy of one's experiences is also a prerequisite for two other unique features of human behaviour, the phenomena we know as religion and science. Some of my scientific colleagues (notably the embryologist

Lewis Wolpert) are, however, inclined to suffer apoplexy when anyone suggests that science and religion are similar phenomena.

In one sense, of course, these colleagues are right: science and religion use radically different methods for making their claims about the world. One is a matter of belief, in which revealed truth holds centre stage as the final arbiter of all disputes, whereas to the other, individual scepticism and the rigorous testing of hypotheses based on logical deduction and reference to empirical evidence are all-important.

But at another level, their apoplexy is premature, for they overlook an important respect in which the two phenomena are virtually identical. Both are attempts to explain the world in which we live. Both serve to give the phenomenal world as we experience it sufficient coherence to enable us to steer a reasonably sensible path through the vagaries of everyday life. The radical difference in the way these two activities work should not obscure the common purpose they serve.

Religions the world over provide security and comfort, a crutch that helps us through the difficult and often dangerous business of daily life. They give us a sense that all is not completely beyond our frail control, that via prayer and ritual we have recourse to mechanisms that will allow us to ensure that life will proceed in a tolerably benign way. In traditional societies, where flood, famine and marauding animals and humans are a constant threat to life and peace, resort to the supernatural may make the difference between sanity and insanity. Having carried out all the necessary rituals, we at least have sufficient sense of certainty to be able to proceed; religion may not entirely prevent the worst from befalling, but it probably provides us with enough confidence and courage to brush off the lesser inconveniences of life that might otherwise have overwhelmed us. In this sense religion is, as Marx famously observed, the opium of the people: it acts just like endogenous opiates, dulling the minor irritations of daily existence just enough to allow life to proceed.

Science too provides us with a framework for existence and allows us to control the world. But the way it does so is, of course, completely different. Science's dramatic success rests not (as some wishfully hope) on arbitrary constructions of reality, but on the careful deduction of hypotheses and their rigorous testing against the events of the real world. Science allows us to have more confidence in its findings because they have to work in the real world. Short of a grand conspiracy theory of science – which would be hard to sustain – it is difficult to imagine how anyone could force the world to produce results that happen to suit the convenience of scientific theories. The real world is simply not that kind of place: it is unyielding, and very unforgiving of incompetents.

The common origins of science and religion lie in a hesitant questioning of why the world is as we find it. The answers they provide may be as different as chalk and cheese, but their function remains the same. And they both depend on the same questioning attitude towards the world. Why is the world as it is? Even to ask that question requires you to be willing to imagine that the world could be other than as it seems. It requires ToM. As such, it is spun off the deep reflexivity of our social behaviour, our ability to understand how an individual's mind can influence his or her actions and how I, in turn, can influence that individual's mind. It requires third-order intensionality at the very least, and quite possibly something beyond that.

If science and religion require fourth-order intensionality, then it is clear why only humans have produced them. Since no non-human animals other than great apes aspire to more than second-order intensionality, none of these species will ever produce science and religion as we know it. But a question mark remains over the great apes. If fourth-order intensionality is essential, then they almost certainly cannot aspire to science or religion. But if third-order suffices, then it is just conceivable that great apes do have science and/or religion.

However, if the apes do have some form of science or religion, it cannot be very sophisticated. Nor will it be a unifying force in their social lives. This is because they do not have language. Language allows us to communicate ideas to each other with an efficiency that is otherwise impossible to match. Without language, each individual has to reinvent the intellectual equivalent of the wheel for itself. We can see and copy tools or someone else's wheel, but religion and science belong to the world of ideas, and we cannot see and copy ideas or concepts in quite the same sense. Without language, we each live in our own separate mental world. *With* language, we can share the worlds inhabited by others. We can discover that other peoples' worlds are not quite the same as ours; that in turn will prompt us to realize that the world can be other than we suppose it to be.

The psychologist David Premack concluded that his star chimpanzee Sarah's mind had been 'upgraded' by being taught a symbolic language. Premack's view seems to be based on the common claim among social linguists and anthropologists that language determines how we think, that without language we cannot have thoughts. In fact, this flies in the face of a great deal of evidence showing that animals do think, that they can develop concepts and all the phenomena we associate with language. It seems more likely that language is parasitic on thought, that it has the kind of grammatical structure we give it (the subject-verb-object form) because that is how we naturally think.

I am not so convinced that Sarah's mind was upgraded merely by the learning of a language: the language did not suddenly create concepts or knowledge that her mind did not previously possess. Rather, Sarah's mind was upgraded by language because language provided her with access to Premack's mind. He was able to pass on to her concepts and ways of looking at things that she might never have thought of on her own. And the emphasis here is very much on the 'might' rather than the 'never'.

Language is thus a crucial factor in the history of ideas. It allows us to build on the knowledge of earlier generations. But it also allows us to exchange knowledge amongst ourselves so that the whole community becomes wrapped up in the same set of beliefs. If chimpanzees have religion, they must have as many religions as there are individual chimpanzees.

CHAPTER 6

Up Through the Mists of Time

Picture the scene some five million years ago. The sunlight dapples the floor of the ancient forest, while the monkeys chatter as they tumble through the treetops on the way from one tree loaded with wild figs to another. On the forest floor there are several species of great ape, not too dissimilar from the chimpanzees and gorillas of today. They travel mostly along the ground, climbing into the trees to forage for fruits and other delicacies.

These apes are the remnants of a family of species that dominated the forests of Africa and Asia for the better part of 10 million years. But times are hard for them. The forests of Africa are contracting under the steady cooling and drying of the world's climate. More and more species are being crammed into a smaller and smaller space. To make matters worse, the monkeys have stolen a march on the apes and are able to outperform them in the ecological race (see Chapter 2). The apes, once the most abundant of the forest primates, are now in decline.

One ape lineage, it seems, eventually began to make more use of the forest edge, venturing further and further out from the safety of the forest to search for food trees that had not already been cleaned out by monkeys. In the woodlands that lie beyond the forest edge, the distance between food trees is greater and the canopy less continuous. You cannot travel, monkey-fashion, from one tree to another along intersecting branches. Instead, it is necessary to descend to the ground and travel overland from one tree to the next.

Stand Tall to Stay Cool

In the less heavily forested woodlands, animals travelling between trees are exposed to more heat from the sun. Peter Wheeler, an ecological physiologist from Liverpool's John Moores University, has studied the heat stress these ancestral apes would have experienced as they moved through the wooded savannahs of Africa. His calculations show that an animal which walks upright receives up to a third less radiant heat from the sun, especially during the middle of the day when the sun is at its hottest. This is simply because less of the body surface is exposed to the direct rays of the sun when standing upright than when walking on all fours. It is a point intuitively obvious to sunbathers: they always lie down to expose as much of the body surface as possible. You'll never get brown quickly standing up.

Moreover, on two legs you benefit from the slight increase in wind speed that occurs above the surface of the earth. Friction from the vegetation and even the ground itself slows the wind down close to the earth's surface in much the same way that a brake acts on a wheel. The increase in wind speed has a significant cooling effect from about three feet above the ground. Large animals of course benefit from this, but smaller animals can benefit too if they stand on their hind legs. Animals about the size of chimpanzees are in the narrow range of body size where standing upright is worthwhile. Smaller species like baboons are not tall enough for standing on two legs to make any difference.

So by standing tall these apes kept cool, which enabled them to travel further into more open habitats in search of food. In the process, another device came into play. With less body surface exposed to the direct rays of the sun, there is less need of the fur that normally serves to keep animals' skin cool. Fur is inert and a good insulator, so the outer tips of the hair can heat up dramatically without the heat being transferred to the body underneath.

Wheeler has argued that nakedness evolved early in the ape lineage that led to modern humans as a way of adding in the extra cooling properties of sweating through bare skin. The upright ape's body was protected from the worst of the sun's rays by its vertical position; consequently, the cooling effect of the wind above the ground-layer vegetation, combined with the cooling effect of evaporation brought on by sweating, made hairlessness a distinct advantage. So we lost our fur coats, retaining them only on those surfaces still exposed to the sun at midday, namely the top of the head and shoulders. Peter Wheeler's careful calculations from the well-established equations for thermal physiology suggest that a hairless, bipedal, sweating hominid could have doubled the distance it travelled on a pint of water compared to a furred quadrupedal one, a saving that would have had enormous advantages for a semi-nomadic hominid out on the open savannahs.

We do not, of course, know exactly when our ancestors lost their fur, because soft tissues and fur are almost never preserved in the fossil record. But we do know that they began to walk bipedally at a very early stage. Two sources of information confirm that for us. One is the shape of the hips and leg-bones of the earliest fossil hominids. The half-skeleton named Lucy that Don Johanson dug out of the Ethiopian Afar desert in 1976 has a well-preserved pelvis and associated leg-bones which clearly prove that this diminutive early hominid from 3.3 million years ago already walked upright. We can tell this from the shape of the pelvis and from the way the knee and hip joints articulate; modern humans have a bowl-shaped pelvis which provides a more stable platform to brace the legs against during walking, whereas apes have a long thin pelvis designed to give more support when climbing. It is clear from the shape of her bones that Lucy's style of walking was not yet fully human: she would have waddled somewhat, rather than walking with the balanced stride so characteristic of modern humans. Moreover, her fingers were longer and more curved than ours and her chest and arms were more strongly built, suggesting that she was still well-adapted to clambering about in trees. But when on the ground, it seems she almost certainly walked upright.

The clinching evidence for this comes from a set of seventy or so footprints preserved under a layer of lava spewed out by a volcano some 3.5 million years ago near a place in northern Tanzania called Laetoli. Here, three sets of tracks closely follow each other, crossing and being crossed by the tracks of antelope and other mammals, across a thirty-yard stretch of what had once been open plain. Walking in the soft lava ash being spewed out by a nearby volcano, the individuals in question left their dramatic imprint on history because a shower of rain shortly afterwards set the lava into concrete. Hidden beneath more layers of ash, what may have been the last actions of this small group were preserved until they were uncovered by the palaeontologist Mary Leakey almost four million years later, in 1978.

There can be no doubt that these footprints were made by a small, bipedal apelike creature about the same height as Lucy. There are no hand-prints such as would be seen when a baboon or chimpanzee walks or runs. Moreover, the big toe is tucked in beside the other toes at the front of the foot just like ours is, rather than being set off at right angles nearer the heel as it is in apes. This was a truly bipedal species that habitually and comfortably walked upright.

Further light has been thrown on the story by new fossil finds in southern Africa. These include a foot on which the big toe is not quite parallel with the other toes, as it is in the modern human foot, although it doesn't stick out quite as much at right angles to the foot as do the big toes of living apes. Dating from around the same time as the Laetoli footprints, this fossil foot suggests an animal that could walk upright but was still at home in the trees.

So it seems certain that an upright stance, and hence probably hairlessness, evolved at a very early stage in our ancestry, while 'we' were still very much apes. It was to be another two million years or more before our brain size would begin to expand significantly beyond that typical of living apes.

Crisis on the Forest Edge

Launching out from the forest into the woodlands that lay beyond provided these early hominids with the advantage of food sources that were less heavily competed for. Most of the species that inhabit the grasslands and woodlands of eastern and southern Africa are grazers or browsers, eating mainly grasses or the leaves of herbs and small bushes. Few species compete for the fruits and seeds that grow on the trees and taller bushes. But nothing comes free in life, and in order to reap the benefits from foraging in this new environment, these ancient apes had to contend with significantly higher predation risks.

As we saw, primates in general exhibit two responses to increased predation: they grow physically bigger and they increase the size of their groups. Our ancestors appear to have done both. Stature increases steadily through time in the fossil record. Little Lucy was barely four feet high when she roamed the savannah woodlands of the Horn of Africa some three million years ago. By 1.75 million years ago, the so-called Narikitome Boy, whose skull and partial skeleton were unearthed by Mary Leakey's son Richard on the shores of Lake Turkana in northern Kenya in 1984, was already pushing five foot three when he died at the age of eleven. Had he lived, he would have been a willowy six foot something as an adult.

If these pre-humans were following the general primate pattern, then it is likely that their group sizes were also increasing steadily through this period in response to the same pressures. How on earth are we to know what groups they went around in? Groups do not leave evidence in the fossil record, and no reputable palaeontologist has ever made any serious claims about group sizes (at least prior to the appearance of permanent camp-sites within the last 100,000 years). It seemed as though we would never know.

Yet our discovery that group size in primates is closely related to neocortex size raised the possibility of being able to estimate, even if with a degree of error, group sizes for fossil species. Being able to do so has other tantalizing prizes to offer. The relationship between group size and grooming time in primates might allow us to solve another elusive question: when did language evolve?

Conventional wisdom offers two main suggestions, both based on very indirect evidence. The archaeologists favoured a relatively recent date around 50,000 years ago when the archaeological record undergoes a dramatic and sudden change known as the Upper Palaeolithic Revolution. At this point there is a marked change in the quality and variety of stone tools. A wider range appears over the succeeding millennia, including awls and punches, and then needles, buttons and clasps. Art objects, such as the exquisite Venus figures and the cave paintings, appear around 30,000 years ago. Burials appear to be organized, with the body placed in a carefully prepared position and often accompanied by objects that might be useful in the afterlife. It all suggests that the people concerned are explaining things to each other, and are able to discuss sophisticated metaphysical concepts such as death and the afterlife.

In contrast, the anatomists favoured an earlier date, perhaps around 250,000 years ago at the latest, associated with the appearance of the first members of our own species, *Homo sapiens*. Their evidence was based mainly on the fact that an asymmetry in the two halves of the brain could be detected at around this point. In modern humans the left hemisphere of the brain, the half where the language centres are located, is larger than the right (for more on this, see the following chapter). They suggested this as clear evidence for the appearance of language.

The disagreement between the archaeologists and the anatomists seemed impossible to resolve, because each side had evidence from its own field to support its views. Leslie Aiello and I thought we might be able to settle this dispute if we could only solve the problem of group sizes in our early ancestors. Our argument went like this.

We knew that no primate spent more than 20 per cent of its day engaged in social grooming. Since modern-day monkeys obviously coped well enough with this (and could probably manage a bit more at a push), the threshold that triggered language evolution must lie somewhere above this. At the same time, we knew it must lie well below the 40 per cent of social time that our equations predicted for modern human groups (see page 77). Somewhere in between, at around 30 per cent of social time perhaps, lay the great Rubicon that precipitated language. But it seemed that we would never know, because we had no way of determining either of the two key variables needed to predict grooming time, neocortex size or group size.

While we were mulling this problem over one evening, we noticed that the neocortex ratio – the proportion of the brain made up of neocortex – in primates (including modern humans and, it later transpired, carnivores too) is directly related to total brain size. While we obviously know nothing at all about the neocortices of fossil hominids, we do have many complete or near-complete fossil skulls from which we can estimate total brain size. Given this, it would be a simple matter to estimate the relative volume of neocortex from the internal dimensions of each skull, and then use this to predict group sizes from the equation we had found between neocortex ratio and group size in primates (see Figure 2, page 63).

The next day was spent in a flurry of calculation, double-checking figures and re-examining the logic of our argument. Then Figure 3 rolled triumphantly off the computer screen onto the printer. This suggests that group size increases rather slowly at first. Moreover, it remains well within the range of group sizes observed in living great apes (especially chimpanzees) until around two million years ago. At this point, a new genus appears in the fossil record, the genus *Homo* to which we modern humans belong. Now, for the first time, group size begins to edge above the upper limits seen in modern primates. From this point on, group size rises exponentially, reaching the 150 that we found in modern humans (see Chapter 4) some time around 100,000 years ago.

The burning issue is: when did group size pass through the critical threshold where language would have become necessary? Looking at the figures for the corresponding grooming times (shown in Figure 4), we concluded that the evidence came down strongly in favour of the earlier of the two dates. By 250,000 years ago, group sizes were already in the region of 120 to 130, and grooming time would have been running at 33 to 35 per cent (well above the critical 30 per cent threshold).

But a closer look at the figures suggests that we might even have to push the date back earlier still. The earliest members of our species appear around 500,000 years ago, and the equations would predict group sizes of 115 to 120 for them, with grooming times of around 30 to 33 per cent. The conclusion seems inescapable: the appearance of our own species, *Homo sapiens*, was marked by the appearance of language.

Our immediate predecessors, the late members of the genus *Homo erectus* – who had relatively smaller brains – may already have begun to run into the same problem. In their case, group sizes were sometimes large enough to nibble at the 30 per cent grooming time threshold. However, most of their groups hovered in the range 100 to 120, with grooming time requirements of

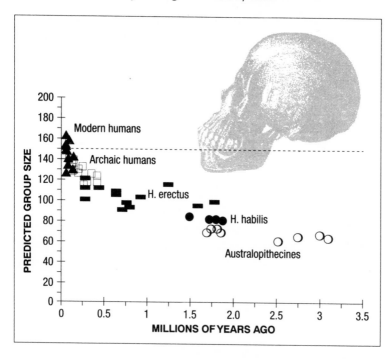

Figure 3. Predicted group size for individual populations of fossil hominids, plotted against their age. The group size is that predicted by the relationship between group size and neocortex size for primates in general (see Figure 2). Five main groups of fossil hominids are represented: the australopithecines (the earliest hominids); *Homo habilis* (the earliest member of our own genus); *Homo erectus* (the first hominid to migrate out of Africa into Europe and Asia); archaic *Homo sapiens* (the earliest members of our own species, including the Neanderthal peoples of Europe and the Near East); and fossil modern *Homo sapiens* (principally Cro-Magnon peoples of Europe and their African relatives). Relative neocortex size was estimated from total brain volume. Each point is the mean for one population (defined as all fossil specimens obtained from the same site dating from within 50,000 years of each other). The horizontal line indicates the group size of 150 predicted for contemporary humans.

about 25 to 30 per cent. Leslie Aiello and I took the view that *Homo erectus* as a species did not have language, though this is probably arguable for the very late members of the species.

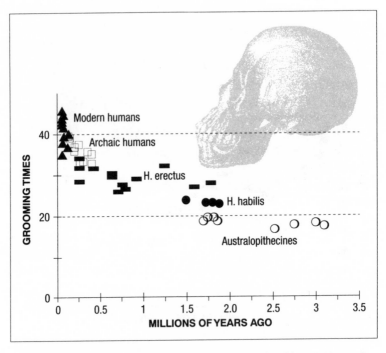

Figure 4. Grooming time predicted for individual fossil hominid populations, plotted against time. Grooming time is that predicted by the
observed relationship between grooming time and group size in living
Old World monkeys and apes, with group size for fossil hominid populations predicted by neocortex size (see Figure 3). The upper horizontal
line indicates the grooming time requirement of 40 per cent predicted for
contemporary humans with their group sizes of 150; the lower line indicates the maximum grooming time observed in any living primate
population (20 per cent in gelada baboons).

However, something else struck us when we looked closely at
the patterns in Figures 3 and 4: there is no obvious jump in
grooming time that would suggest the crossing of a major
Rubicon. Had the critical point coincided with the appearance of
the first members of the genus *Homo* at around 2.5 million years
ago, we might have been able to conclude that language appeared
rather suddenly. But there is no such 'big bang' around half a million years ago, which suggests that, rather than evolving as the

result of some dramatic new mutation, language emerged slowly over a long period of time.

It seems that language as we know it evolved in at least three stages, becoming progressively more complex as the demands of larger group sizes became more pressing. We suggest that it has its origins in the conventional contact calls so characteristic of the advanced Old World monkeys and apes. In these species, as we noted in Chapter 3, contact calling functions as a kind of grooming-at-a-distance. As time-budgets became increasingly squeezed, the animals would have kept up a steady flow of vocal chatter. Its content would have been zero, rather along the lines of those formulaic greetings so common in our own conversations. Remember the hackneyed 'Do you come here often?' This is not a question requiring a carefully thought-out answer. It's an opening gambit, a tentative searching for a reply along the lines of, 'Yes, I'd be very happy to spend the next half-hour in your company, thanks ...'

We suggested that, as group sizes began to drift upwards from the numbers to which apes are currently limited, vocal grooming began increasingly to supplement physical grooming. This process would have begun around two million years ago with the appearance of *Homo erectus*. More and more emphasis was being placed on vocal as opposed to physical grooming as a bonding mechanism.

Eventually, even this form of communication would have exhausted its capacity to bond groups. A more efficient mechanism for bonding was needed to allow group size to continue its upward drift. At this point, the vocalizations began to acquire meaning. But the content was largely social: gossip had arrived.

This need not have involved any dramatic change, for as the studies by Seyfarth and Cheney have shown, primate vocalizations are already capable of conveying a great deal of social information and commentary. The pieces of the jigsaw were already there: all they needed was to be organized into a coherent system. The continuing drive for ever-larger groups provided the necessary kick at the right moment.

In effect, humans were now exploiting the greater efficiency of language as a bonding mechanism to allow themselves to live in

larger groups for the same investment in social time. This suggestion is borne out by the fact that modern hunter-gatherers seem to
devote about the same proportion of their day to social interaction as the modern gelada (the species that holds the record for
time spent grooming among non-human primates). A study of the
Kapanora tribe in New Guinea, for example, found that men
spent an average of 3.5 hours and women around 2.7 hours in
social interaction during a typical 12-hour day. In other words,
they spend on average a quarter of the day socializing, compared
to a figure of 20 per cent for gelada.

Not until very much later, perhaps as much as 400,000 years later, do we see any evidence for the appearance of symbolic language: language capable of making reference to abstract concepts.
At this point in the fossil record, we see a sudden change in the
style, quality and variety of stone tools. The previous two million
years had seen almost no development in the types of stone tool in
common use. They were limited both in the nature and complexity
of their design, and hence in the skills required to fashion them;
most are crude choppers and handaxes of little artistic merit. Then
quite suddenly the picture changes. Tools become more delicate
and more finely prepared. Substances such as red ochre appear in
the fossil record for southern Africa with evidence that they have
been ground and rubbed, suggesting they may have been used in
tanning skins or manufacturing body paints. Here then are the first
hints of ritual. The cultural revolution had arrived.

This seems to resolve another controversy that has rumbled on
over the years: did the Neanderthals have language? The
Neanderthals occupied Europe throughout the Ice Ages, from
around 120,000 years ago. They suddenly disappeared about
30,000 years ago after a period of coexistence with the ancestors
of modern humans, the Cro-Magnon people, who arrived from
Africa around 50,000 years ago. The anatomist Philip Lieberman
has always insisted that the Neanderthals did not have language,
on the grounds that their larynx (the top of the tube from the
lungs) was too high in the throat to produce human vowel sounds.
The Neanderthal larynx, he suggested, was in much the same
position as the chimpanzee's, and they certainly cannot produce

such sounds. Unable to communicate with each other in anything more than grunts and screams, the dumb Neanderthals were over-run by the tall, slim modern humans from Africa with their sophisticated culture and their language.

But Lieberman's claims have been thrown into doubt by the dis-covery of an almost complete Neanderthal skeleton in Israel with its hyoid – the tiny bone that supports the larynx and the base of the tongue – still in place. The Neanderthal's larynx appears to have been roughly where ours is, low enough in the throat to pro-duce the full range of speech sounds. Anatomically, it looks as though they did have language. Our analyses would agree with this. With brains that were if anything slightly larger than those of modern humans (the Neanderthals were larger and much stronger than we are), they must have had groups of the same size as other modern humans. Grooming times would have been well beyond the sustainable limits. So if they were not using their brains to maintain human-sized groups, what on earth were they doing with them? It must have been something quite different from all other living primates, including modern humans.

If the Neanderthals became extinct at the hands of the invaders from Africa, it was not because they lacked language but because they lacked the sophisticated culture and social behaviour of our African ancestors. Not only were the stone tools and artefacts of the Cro-Magnons very much more delicate and complex than those of the Neanderthals, but there is also evidence that the Cro-Magnon people traded shells as well as flint and other stones over very wide areas. They clearly had much more sophisticated and widespread social networks than the Neanderthals. Recent research on the many cave sites in Israel suggests that the Neanderthals may have been more sedentary in their habits, whereas the modern humans who lived in the area during the same period were more nomadic and followed the changing distri-bution of resources more closely. The moderns thus appear to have been ecologically more flexible.

The fate of the Neanderthals bears an uncanny resemblance to the fate of the American Indians and the Australian Aborigines at the hands of the later Europeans, invaders who could draw on a larger and more widely distributed political and military power base. Old habits, it seems, die hard – though in the latter case it is worth reminding ourselves that all the parties involved (Europeans, Americans and Australians alike) were the direct descendants of those Cro-Magnon invaders from Africa.

We are left with one last puzzling question: what drove the increase in group size? The short answer is, we don't know. But we can hazard some guesses. The conventional wisdom on primates is that there are only two likely pressures that select for large group sizes: one is predation risk, the other the need to defend food sources. But if baboons can cope with groups of around 50, then it is difficult to see why the later humans and their immediate predecessors should have needed groups that were nearly three times bigger. The fact that they were larger than baboons (and carried passable defensive weapons in their hands) should have meant that they could get by with smaller groups in the same habitats. And indeed, this is pretty much what you see. Baboons typically live in groups of 50 to 60 in woodland habitats, while hunter-gatherers in eastern and southern Africa typically live in temporary camps of around 30 to 35.

Of course, our ancestors may have invaded even more open habitats than those currently occupied by the woodland-based baboons and the forest-loving chimps, and so might have needed larger groups to offset the greater predation risk. Some evidence to support this suggestion is provided by the modern gelada. This species lives in very open habitats where they have few trees in which to escape from predators; they live in the largest naturally occurring groups of all primates (typically 100 to 250 animals). Moreover, the size of their groups correlates with the predator risk in the habitat, being larger in those habitats that provide them with fewer safe refuges.

A second possibility is that it was other human groups rather than conventional predators that posed the threat. This could have been in the form either of raiding (for example, for women, as happens among some hunter-gatherers today) or of competition for food and/or water resources. Increases in both group size and individual stature could just as easily be interpreted as a response to these kinds of problems as to predation risk. It is just the kind of situation in which an arms race occurs: the raiders form bigger groups to raid more successfully, so you need to form even bigger groups to protect yourself, so the raiders need to form bigger groups still, and so on until ecological constraints impose a limit (perhaps in terms of the number of individuals that can be fed).

A third possibility may stem from the fact that early in the second phase of human evolution (following the appearance of *Homo erectus* two million years ago), a dramatic change in ecological behaviour occurred: our ancestors became nomadic. They crossed the Arabian land bridge into Asia for the first time and within a few hundred thousand years had reached China and the islands off the south-eastern coast of Asia.

Nomadism on this scale suggests that the groups foraged over large ranges, but were not afraid to step into the unknown beyond their everyday boundaries in search of new food sources. Animals face two problems in these situations. One is that they are unfamiliar with the lie of the land and don't know where the safe refuges or the good feeding places are. The Swiss biologists Hans Sigg and the late Alex Stolba showed that when the occasionally nomadic hamadryas baboons are travelling outside their normal territory they are significantly less likely to find water-holes and food trees than the groups resident in the areas concerned. The other is that resident groups may actively exclude them from access to such essential resources (and water-holes may be absolutely crucial in hot savannah habitats). Migrants are always at a disadvantage.

Establishing reciprocal alliances with neighbouring groups may be the only way to solve this problem. In effect, a set of neighbouring groups would start to act co-operatively to share their water-holes or other key resources. The result would be an alliance of loosely federated groups that could come and go, merge and split up, as the mood dictated. And this is exactly what one sees in modern hunter-gatherers.

The !Kung San[1] of the Kalahari, for example, live in communi-

---

1. The !Kung San speak a so-called 'click' language in which click-like sounds made by the tongue and lips form part of the range of vowels and consonants in the language. The '!' is how one such sound is represented.

ties of 100 to 200 individuals, each centred on a set of permanent water-holes that they 'own' or have rights over. The community itself, however, rarely appears as a single group. Rather, its members forage in small groups of 25 to 40 (typically four to six families). The permanent water-holes are the community's life-blood, providing a safe retreat in times of drought when temporary water-holes dry up.

This is known as a fission-fusion social system, because its members are constantly coming and going. Chimpanzees share this characteristic with us, except that their groups are smaller. Chimpanzee communities are around 55, while the foraging parties in the forested habitats they occupy often number only three to five individuals. But the fact that we share this characteristic with chimpanzees does suggest that the precursors for modern human societies of this kind were already established during the earliest phases of our history.

Each of these three possibilities has evidence in its favour. If I had to guess which one was most likely to be right, I think it would have to be the last, in part because it fits modern hunter-gatherer behaviour patterns and in part because we seem to share this pattern with the chimpanzees.

Testing the Hypothesis

If language evolved to facilitate the bonding of larger groups, then we should be able to show that it has design features that would achieve this. One is that conversation groups should be proportionately larger than conventional primate grooming cliques. Another is that conversation time should be predominantly devoted to the exchange of social information. In one sense at least, the latter would be a strong test of the hypothesis, because conventional wisdom has it that language exists to facilitate the exchange of information about the world in which we live – the view that we spend our time discussing the bison down by the lake.

In order to test these predictions, my students and I sampled conversation groups in various locations. Our only prerequisite was that the people concerned should be engaged in relaxed social interaction with friends. We wanted to avoid formal situations, where the rules of conversation are often deliberately constrained. We sampled conversations in university cafeterias, at public receptions, during fire practices (while everyone was waiting for the all-clear to go back inside the evacuated building), in trains and in bars.

The first thing we found was that conversation groups are not infinitely large. In fact, there appears to be a decisive upper limit of about four on the number of individuals who can be involved in a conversation. The next time you are at a social gathering such as a reception or a party, take a look around you. You will see that conversations begin when two or three individuals start talking to each other. In due course, other individuals will join them one by one. As each does so, the speaker and the listeners try to involve them in the conversation, directing comments to them or simply moving to allow them to join the circle. However, when the group reaches five people, things start to go wrong. The group becomes unstable: despite all efforts (and groups often do try), it proves impossible to retain the attention of all the members. Instead, two individuals will start talking to each other, setting up a rival conversation within the group. Eventually, they will break away to start a new conversation group. This is a remarkably robust feature of human conversational behaviour, and I guarantee that you will see it if you spend a few minutes watching people in social settings.

Since there is only ever one speaker at a time (aside from momentary overlaps or attempts to butt in), this limit of four on the size of conversation groups means there are three listeners. That is a particularly interesting number, because it is three times the number of parties involved in a conventional grooming interaction – which always consist of just one groomer and one 'groomee'. The largest mean group size observed in any primate species is about 55, in chimpanzees (remember that these are *mean* group sizes, not the largest groups ever observed in a given species). This *may* be an extraordinary coincidence, but the predicted (and observed) size of modern human groups, 150 individuals, is almost exactly three times larger. In other words, the ratio between group sizes is exactly the same as the ratio between the number of individuals you can interact with at any one time: one for grooming monkeys, three for talking humans. My view is that human groups are three times larger than those of chimpanzees precisely because humans can reach three times as many social contacts as chimps for a given amount of social effort.

The story became even more intriguing when I discovered that the size limit on human conversation groups seems not to be an accident of social rules. Rather, it turns out to be a consequence of limitations in our auditory machinery. Our ability to hear what is being said is just good enough to allow us to operate in conversation groups of this size. During the 1950s and 1960s the effect that noisy environments had on the detectability of speech was studied in considerable depth, and two facts of particular interest to us emerged.

One is that, when the distance between speaker and listener is more than about two feet, the listener finds it increasingly difficult to make out what the speaker is saying. Calculations based on the rate at which speech discrimination falls off with increasing distance between speaker and listener suggests that, under minimal noise conditions, there is an absolute limit of about five feet beyond which the listener simply cannot hear enough of what the speaker is saying. Beyond this distance, the speaker has to shout. In fact, Dan Nettle, one of my students, has shown that the loudness with which cultures habitually speak is inversely relalted to the number of vowels in the language, and hence to the ease with which the language's sounds can be discriminated. Cultures which abhor excessively close contact tend to shout at each other and have vowel sounds that are easily distinguished from one another.

Even allowing a minimal shoulder-to-shoulder distance of six inches, a circle five foot in diameter would place an upper limit of about seven on the number of people who can hear what a speaker is saying. However, as background noise increases, so the minimum tolerable distance between speaker and hearer shrinks and the number of people you can get into a circle decreases proportionately. With background noise levels typical of a city street or a busy office, the limit on the group size is about five. At very noisy cocktail parties, it sinks to two (and even then it can be an effort).

The second prediction made by the hypothesis of the social language was that people's conversations should be dominated by social topics. We have listened to conversations up and down the length of England, sampling individuals of different ages and social backgrounds. The technique was very simple. Every thirty seconds we simply asked, 'What is he/she talking about *now*?' Our results consistently yield the same pattern: about two-thirds of conversation time is devoted to social topics. These include discussions of personal relationships, personal likes and dislikes, personal experiences, the behaviour of other people, and similar topics. No other individual topic accounted for more than 10 per cent of total conversation time, and most rated only two or three per cent. These included all the topics you might consider to be of great moment in our intellectual lives, namely politics, religion, ethics, culture and work. Even sport and leisure barely managed to rustle up a score of 10 per cent between them.

After we had completed these studies, we discovered that other scientists had been sampling conversations. Nicholas Emler, a psychologist at Oxford University, has also been particularly interested in gossip and its uses. After listening in to conversations in Scotland while he was based at Dundee University, he too came up with a figure of about 60 to 70 per cent devoted to social topics. He concluded that one of the most important things gossip allows you to do is to keep track of (and of course influence) other people's reputations as well as your own. Gossip, in his view, is all about the management of reputation.

Taken together, these observations provide strong support for the suggestion that language evolved to facilitate the bonding of social groups, and that it mainly achieves this aim by permitting the exchange of socially relevant information.

The Expensive Tissue Hypothesis

The evolution of large brains in the ancestral hominids, and the language facility that emerged from this, raises some fundamental questions. Brain tissue is by far the most 'expensive' tissue in the body. It differs from other body tissues in the extraordinarily high levels of activity it maintains even during the resting state. Nerve cells work by maintaining an electric potential across the cell membrane by forcing sodium ions (free atoms) out and potassium ions in. Allowing the floodgates to open creates the electrical surge that we identify with a nerve firing. Since the ions naturally want to move from one side of the membrane where their concentration is high to the side where it is low, it is necessary to have a mechanism that pumps ions in the opposite direction in order to keep the nerve ready to fire. A great deal of energy has to be expended by the ionic pumps to keep the electrical potential in the axon at the right level in readiness for action.

In addition, the neurotransmitters that facilitate the transmission of electrical discharges between one neuron and the next are very expensive to manufacture, and considerable energy has to be expended in replacing these every time an axon fires. Neural tissue is, in fact, about ten times more expensive to maintain per kilogram than the rest of the body's tissues, consuming about 20 per cent of the body's total output of energy while accounting for only 2 per cent of its weight.

This creates something of a problem. As we saw in Chapter 4, modern humans have brains about nine times larger than you would expect for a mammal of our body weight, and about six times larger than for a primate of our body size. Yet gross energy production is directly related to a body's size. So even though we have a brain that is much larger (and hence more expensive) than is typical of other primates, the total energy output of our bodies is just what would be expected for a typical mammal of our size. How then do we manage to maintain a larger brain when we don't have any extra energy to fuel it?

The answer, as Leslie Aiello and Peter Wheeler have pointed out recently, is that we have a much smaller gut than would be expected for a mammal of our size. When they looked at the energy consumption of different parts of the body, they discovered that, rather surprisingly, the brain, the heart, the kidneys, the liver and the gut between them account for 85 to 90 per cent of the body's total energy consumption in mammals. Hence to increase the size of the brain, the extra energy required to fuel it must come from one of these organs.

The problem is that the heart, kidneys and liver are related rather closely to body size in mammals, for the obvious reason that they are responsible for maintaining healthy tissue activity. A smaller heart means that less blood is pumped round the system, and therefore the muscles cannot work so hard; a smaller kidney or liver means that the blood cannot be purified so effectively, and thus carries less energy to the muscles (and an increased risk of the organism being poisoned through failure to extract toxins from the blood). In short, if you want a large brain, the only place you can really afford to take the extra energy from is the gut.

The catch, of course, is that you cannot reduce the size of the gut without reducing the rate at which energy is absorbed into the bloodstream from the food you ingest. Catch 22. The rate at which the gut extracts energy from food is directly related to its area – which is, of course, in turn related to its overall volume. It seems that gut size ultimately places a constraint on brain size. If you want to have a bigger brain, you have to grow bigger in order to accommodate a bigger gut. The lesson here is that small monkeys can never be very smart – never mind evolve language – because they do not have guts large enough to support the extra neural activity of a bigger brain. They simply don't have enough scope for savings of scale.

But there is one way out of this bind, and it is precisely the way the ancestral hominids found. You can reduce your gut size without compromising your energy intake by eating foods that are either richer in nutrients or have their nutrients in a more easily assimilated form. That way, the gut has to work less hard to extract the same quantity of nutrients.

Most primates are either leaf-eaters or fruit-eaters, although many of the small nocturnal species are insectivorous. Although insects are a very rich source of energy, only very small animals can support themselves on such a diet, because insects are hard to catch and even the largest of them come in rather small packets. Of the two diets available to the medium to large primates, the less nutrient-rich – or at least, the one from which it is harder to extract nutrients – is the leaf-based diet. The cell walls of most leaves mainly consist of various kinds of cellulose which mammals cannot easily digest.

Species with a leaf-based diet cope with the problem by using bacteria to ferment the leafy matter; the host then consumes the bacteria further on in the gut, so absorbing the nutrients from the cell indirectly. This is basically what cows, antelope and other ruminant mammals do: rumination is a period of fermentation during which the bacteria digest the leaf cells. Some primates like the colobus and langur monkeys of the Old World and the howler monkeys of the New World behave in a similar way (though without ruminating or chewing the cud). There are severe costs to this strategy, though. In order to ferment the leaves, you need a big vat: all leaf-eaters amongst the mammals have very big guts – large stomachs in which to ferment the leaves, and large intestines to provide absorbent surfaces for extracting the nutrients from bacteria). Leaf-eating is basically incompatible with a small gut.

Fermenters face other problems that have important implications for our theory. A fermenting strategy requires time for the bacteria to do their work. It is also a very hot strategy, because the fermentation process generates a great deal of heat, raising the animal's core temperature. But if the body temperature rises, the bacteria cannot ferment effectively. So once they have filled up their stomachs, fermenters have to rest for a while to digest the food particles properly and create some space before they can start feeding again. If you watch a herd of cows in a field for any length of time, you will see this neatly demonstrated. After feeding for a couple of hours, they will all go and settle down in a corner to chew the cud. A couple of hours later, having cleared some space in their guts, they all set about feeding again. And so on throughout the day and night, alternating between feeding and ruminating on a roughly four-hour cycle.

Among primates too there is a positive relationship between the proportion of leaf in the diet and the amount of time devoted to resting. Leaf-eating colobine and howler monkeys can spend as much as 70 to 80 per cent of their available daytime resting; species such as baboons and chimpanzees that feed on more energy-rich fruit-based diets spend as little as 10 to 20 per cent of their time resting. In consequence, the amount of time leaf-eaters have available for social activity is severely curtailed (two to five per cent of their time) compared with fruit-eaters, who may spend as much as 15 per cent of their time grooming. So even if their brains were big enough to support large groups, colobines and howlers would not have sufficient time to spare for grooming to be able to keep groups of any size together.

All the large social primates are fruit-eaters in one form or another. Fruits, seeds and tubers (the underground storage organs of certain plants) are the most energy-rich of all vegetable foods, and their energy is in the form most accessible to primates. As fruit-eating apes, the ancestral hominids could not have significantly improved their diet as a way to reduce their gut size. Only one source of food available to them was more nutrient-rich, and that was meat. Flesh is energy-rich, and the energy is in a form particularly easy to absorb during digestion. As a result, carnivores have rather small guts for their body size. By switching to a meat diet, the ancestral hominids were able to make significant savings in gut volume without sacrificing any of their energy intake.

The initial increase in brain size about two million years ago seems to correlate with a shift from a predominantly vegetation-based diet in the australopithecines (the early members of our lineage) to a diet with a significantly larger meat component in *Homo*. Most of this meat probably came from scavenging the kills of other carnivores and opportunistic killing of birds, reptiles and baby mammals rather than from full-scale hunting. However, the sudden increase in the rate of brain expansion from around half a million years ago seems to coincide with the beginnings of organized hunting. From here on, meat becomes increasingly important in the diet. The living floors of fossil sites in Africa and Eurasia start to contain the bones of large numbers of mammals, including other primates. The bones often bear cut-marks, suggesting deliberate dismemberment. Some species, like the now-extinct giant gelada at the Olorgasaillie site in southern Kenya, appear in such numbers as to suggest that hominid hunting may have been a contributory factor in their subsequent extinction.

In short, Aiello and Wheeler argue that our super-large brains were possible only because we went in for meat-eating in a big way. That allowed us to live in larger groups, and that in turn allowed us to range more widely, colonizing the rest of the world in a series of migrations that fanned outwards from the ancestral homelands in eastern and southern Africa like the ripples on a pond.

The Baby Needs its Bathwater

There were, however, hidden costs to developing large brains. If we plot the brain sizes of different species against their gestation periods in what is usually known as a 'mouse-to-elephant' curve, we find that the two variables are closely related. Mice, with their tiny brains weighing only tens of grams, have a three-week pregnancy, whereas elephants, with their human-sized brains, have a 21-month pregnancy. The fact that all species of mammals lie close to the line between these two extremes suggests that brain tissue is laid down at a constant rate during pregnancy, at a rate determined by the mother's ability to channel spare energy into her foetus. In effect, it is the size of the baby's brain that determines the length of pregnancy, and all species give birth when brain growth is more or less complete. Baby primates, for example, are born when their brains reach full size, with relatively little growth occurring after birth.

Humans, however, are an exception. A baby human is born when its brain is less than one-third its final size. The rest of its brain development continues over the first year of life. In fact, if we calculate the equivalent gestation period for a conventional mammal of our brain size, we arrive at a mind-boggling 21-month pregnancy. This corresponds exactly with the time it takes the human baby's brain to complete its growth: nine months gestation in the womb plus an additional 12 months after birth.

One consequence of this is that human babies are born premature and incapable of fending for themselves. A baby monkey or ape is capable of walking within hours of its birth. Within a matter of weeks it is a competent member of its social group. The average human baby, by contrast, can barely muster even a gurgle to keep its long-suffering parents happy. By its first birthday, however, its brain is sufficiently developed to allow it to learn to walk and begin the serious business of living.

Our premature births became necessary because the human brain size took off at a time when we were actually getting smaller. We needed larger brains to cope with our bigger groups, but we were getting slimmer and shorter in response to other ecological pressures. This wouldn't have been a problem were it not for the fact that the birth canal through which the infant passes as it is being born was only increasing as the square of body length whereas the brain size was increasing as the cube. Trying to squeeze an ever bigger head through what was becoming an ever smaller hole creates inevitable problems. Something had to give.

What gave was the timing of birth. Instead of giving birth to infants whose brains had completed their growth (as is typical of most other mammals), we compromised and gave birth at the earliest possible moment at which the baby could survive, and allowed it to complete its brain growth outside the womb. We give birth to appallingly premature babies. And this is why true premature babies, those born after only six or seven months of pregnancy, have such a difficult time; they really are teetering on the brink of survival – because even when they are on time, human babies are premature.

The costs of growing a large brain were thus considerable for our ancestors. The whole business of successful reproduction, of investment in one's offspring, became drawn out and magnified. The human child takes proportionately longer than the 5 to 10 years typical of monkeys and apes to absorb all the necessary information and experiences into that little brain to allow it to cope with the social world into which it has been born. In humans, the learning process extends over a 15 to 20-year period. In effect, as brain size increases, so everything from pregnancy and the age at weaning to sexual maturation and the business of reproduction gets slowed down and strung out over a progressively longer period.

An increased parental role for males may well have become

essential, because the females (can we now call them women?) could not bear the full costs of child care while foraging on their own. In other words, it must have been at this point that the unusually intense pair-bonding that occurs between human males and females first evolved.

This is certainly implied by the reduced sexual dimorphism in body size. Throughout most of our early history, during the australopithecine phase, males were substantially larger – in some cases as much as 50 per cent larger – than females. In mammals, striking sexual dimorphism is invariably associated with harem-like mating systems, where a handful of powerful males share all the females between them. The reduced sexual dimorphism in the later hominids, where males are only 10 to 20 per cent heavier than females, suggests that females were shared more evenly among the males (though some males still did better than the rest).

One particularly reliable index of the mating system in primates is the relative size of the male canine teeth compared to those of the females. Species in which the males defend large harems, or compete in open arena to mate with individual females on a promiscuous basis, tend to show marked differences between the size of male and female canines, since these are the main weapons used in fights between males over access to females. In contrast, sexual dimorphism in the canine teeth is negligible (or even favours larger female canines) in lifelong monogamists like the gibbons. Among the australopithecines and early *Homo* the dimorphism is considerable, with males having canines that, relative to body size, are about 25 per cent larger than those of females, about the same relative difference as in the highly promiscuous chimpanzees. Canine dimorphism, however, appears to have declined steadily during the last two million years, reaching a low point 50,000 years ago when male canines were only about 10 per cent larger than those of females. This implies a shift over time from strongly polygamous mating systems towards mild polygamy.

This doesn't by any stretch of the imagination mean that monogamy had evolved: despite the intensity of human pair-bonding, there is no anatomical evidence for monogamy. But it does suggest that harem groups were very much smaller, with perhaps only two females attached to a given male. Many males would have had only a single female. This is what we would expect if females were demanding increased provisions from the males whose children they were bearing. If the experience of modern hunter-gatherers is anything to go by, it is unlikely that a male could have provided sufficient meat for more than two females and their offspring. This would at least explain the incongruity between the anatomical evidence for polygamy and the intensity of pair-bonding (with its implications of monogamy) in modern humans.

CHAPTER 7

First Words

The evolution of language raises two fundamental and related questions. First, what form did language (speech?) take when it first evolved? Second, why, and when, did what was presumably once a single language diversify to give us the 5000 or so languages we now have? I am going to try to answer the first question in this chapter; I will address the second question in the following chapter.

Since we are dealing with events in the distant past that leave no fossil record, my answers will necessarily be fragmentary. However, as I shall try to show, we may at last be in a better position than before to provide some serious answers to these questions.

Irrespective of *when* language first evolved, the puzzle is how non-linguistic forms of communication gave rise to a linguistic form. How did non-language transform itself into language? What did those first words sound like? Would we recognize them today as a human language, complete with grammar and all the other accoutrements of modern conversation? And last, but perhaps not least, who spoke them?

Opinions differ about what the earliest forms of language might have been. One school of thought says that it arose from gestures, while another argues that it came from monkey-like vocalizations; yet another suggests that it arose from song. Each hypothesis is supported by enough evidence to make it seem plausible.

It is convenient to start by considering the gestural theories of language origins, partly because they are still widely held and partly because they inadvertently raise issues that will turn out to be important to our story. I will deal with the song-based theories in the following section.

Gestures on the Wind

One version of the gestural theory is based on the observation that the fine motor control used in both speech and aimed throwing tends to be located in the same half of the brain, the left hemisphere in most people. Speaking requires very fine motor control of the lips, tongue, vocal cords and chest, all of which have to be integrated in just the right sequence to produce a particular sound. Just try saying the sound *a:* as in *hay* – normally produced with the corners of the mouth pulled right back so that the lips are narrowly parted – but with the lips rounded and pouted as they would be when producing the sound *oh:*. The result is recognizable as an *a:* sound, but only just; it comes out more like a strangled *oi:* than anything else. In addition to fine motor control, speech also requires precise control of breathing so that air is released from the lungs in just the right quantity and with just the right amount of force (think about the difference between explosive sounds like *b* or *p* and softer ones like *e* or *c*.)

One important anatomical change was necessary for this: the shift from the dog-like chest shape typical of all monkeys to the flattened chest characteristic of apes. The way a monkey's shoulder girdle is attached to its rib cage constricts the frequency with which it can breathe. The scapula (the flat plate of bone that provides the anchor and pivot point for each arm) lies on the side of the rib cage in most mammals, including all the monkeys. This allows the arms to move backwards and forwards when walking or running. The problem is that when the body's weight is on the arms, it restricts the chest's ability to expand and contract during breathing: as a result, monkeys can breathe once, and once only, with each stride.

When the apes adopted a climbing life-style – in which the arms are used above the head to pull the body vertically upwards, while the feet are braced against the tree trunk – the monkey-like rib cage of their ancestors was a serious impediment. Monkeys cannot swing their arms in a circle; the position of the shoulder bones blocks the movement of the arms. To allow them to reach up above their heads during vertical climbing, the ape's chest had to be flattened. With the scapula moved around to the back of the rib cage, the arm joint could be positioned on the outer edge of the chest. With this arrangement, apes could now swing their arms in a complete circle at the shoulder. This is why gibbons (the 'lesser' apes) can brachiate (that is, swing from tree to tree using the body as a pendulum), the great apes can haul themselves up tree trunks and we can play baseball, while the monkeys can do none of these things.

This flattened chest, in addition to preparing the way for the evolution of bipedal walking in our ancestors (it helps to keep our centre of gravity over the feet when we are standing and walking), also freed the breathing apparatus from the constrictions suffered by the monkeys. We now breathe as often as we like, irrespective of what our arms are doing. We can now speak without interruption even when active.

This small but important change in anatomy prepared the way for something else: aimed throwing. Although other apes and monkeys throw things, their accuracy is not always impressive. Only modern humans can throw a cricket ball from the outfield with the intention of hitting the wicket (or at least getting it into the wicket-keeper's hands, given a modicum of gymnastic ability on the latter's part). Yet in terms of sheer arm power, the average chimpanzee could easily out-throw any Olympic field sports champion. Fortunately for all the budding Tessa Sandersons, no ape will ever stand much of a chance on the sports field because they lack the fine motor control required to throw a javelin any distance.

Aimed throwing is clearly important for hunting, so one obvious conclusion is that language evolved on the back of throwing. The fine motor control needed for aimed throwing, so the argument runs, provided us with the neural machinery for fine motor control of the organs of speech. The sensory and motor control nerves from one side of the body cross over to the other side of the brain (the right side of the body is controlled by the left hemisphere of the brain, and the left side by the right hemisphere), and most people throw right-handed; it follows that motor control for throwing is located in the brain's left hemisphere, and hence that speech control will be located there too (as indeed in most people it is).

There are, however, a number of problems with this suggestion. For a start, language involves conceptual thinking of a very different order to that needed for throwing. Secondly, it is difficult to see how a gestural language of any complexity could get going.

We use gestures very effectively to give commands (beckoning to indicate 'come here'), to draw someone's attention to something (pointing), or to emphasize a point (stabbing the air). We use them to express anger (fist-shaking) or submission (hands clasped in petition or held up in surrender), or to indicate friendship (waving or handshaking). But we never use gestures to express abstract concepts, to indicate place or time other than the present, or to make plans for the future. If we could, charades would be a pointless game. (OK, it *is* a pointless game, but the difficulty of expressing concepts in gestural form taxes our communication skills to the point of absurdity, so making it amusing as a parlour game.) More importantly perhaps, we don't use gestures to discuss other people's behaviour (other than through simple comments such as raised eyebrows). In other words, we do not use gestural forms of communication to express much more than the kind of information about emotional states that monkeys and apes competently express using vocalizations (and sometimes gestures too, of course!).

But the real problem with a gestural theory for the origins of language is simply its impracticality: you have to be in close visual contact with the person you're talking to. As the children of deaf parents quickly learn, you can cuss and swear all you like at your parents so long as their backs are turned. More significantly perhaps, darkness occupies exactly half the day in the tropics. Those early linguistic humans would have had to sit in self-imposed 'silence' from dusk until dawn. Unable to tell stories of the old times, let alone argue about where to hunt the next day, the only possible evening pastimes they could have enjoyed were grooming and sex.

Speech, on the other hand, frees us from these constraints. We can reminisce and tell stories over the dying embers of the fire; we can shout instructions or make enquiries over distances of half a kilometre or more, even when we cannot actually see the person with whom we are conversing.

There's another question no one ever thinks of asking: why should the control of throwing have centred on the left side of the brain? Why not the right (with left-handed throwing)? The only explanation is the rather unsatisfactory one that it was an accident of history. Now accidents of history do occur in biology; which genetic sex has the 'female' (or egg-producing) role in reproduction is a classic case.[1] Brain asymmetry is more tricky, because it is not obvious why it had to be one hemisphere rather than the other: why couldn't humans have evolved into ambidextrous throwers, like all other primates?

The answer I want to propose (and it is really no more than a speculation) is that the right hemisphere was already fully preoccupied doing something else much more important. Speech, when it evolved, localized in the left hemisphere because there was more free space there. And once speech had evolved, the fine motor control for throwing also centred there, either for the same reason or because the left hemisphere was already beginning to specialize in conscious thought, which is needed for accurate aimed throwing. In other words, the sequence of events was precisely the reverse of that proposed by all gestural theories.

I suggest this for a very simple reason: we now know that the right hemisphere is specialized for the processing of emotional information. There is evidence to show that emotional cues are detected faster when they are on the left side of the visual field (transmitted across to the right side of the brain[2]) than on the right side. This trait is widespread in the animal kingdom, and

1. It is the XX-chromosomed sex in mammals, but the XY-chromosomed sex in birds and butterflies – the result of an entirely random decision on three quite separate occasions as to which sex should have the eggs and which the sperm.
2. Unlike all other systems in the body, vision is split: the nerves from the left half of each eye are fed across to the right side of the brain, and those from the right half are fed across to the left side.

appears to originate in a very ancient tendency for greater sensitivity to visual cues to have evolved on one side. Fossil trilobites from 250 million years ago, for example, tend to have more scars on the right side, suggesting that pursuing predators attacked them more frequently from the (predator's) left-hand side. Fossil dolphin skeletons from 20 million years ago show the same pattern of shark-tooth damage on the right-hand side, again suggesting that the predator kept the prey in its left visual field as it chased it.

Evidence from living species suggests that this tendency became elaborated into a generally greater sensitivity to visual and emotional cues on the left side of the body. Julia Casperd and I have shown that gelada males tend to keep opponents on the left side of the visual field during fights. During the early phases of these encounters, the animals exchange threats using facial signals; if neither of them will submit, the confrontation will eventually escalate into a physical attack. It is obviously important for the males to monitor their opponent to pick up any inadvertent hint it gives about its true intentions. Is it bluffing when it makes heavy threats? Are the eyes that momentarily flicker away betraying a reluctance to press home an attack if pushed to the brink? The greater sensitivity of the right hemisphere means that keeping the opponent in the left visual field (so its image falls on the right half of each retina[3]) ensures that the subtler cues are picked up.

We humans do this too. Have you ever noticed that when people are photographed, they tend to turn the head so that the left side of their face is directed at the camera? Take a look at your family photo album. Only in very formally posed shots such as team photographs do people stare straight at the camera; in informally posed photos, taken when people know they are being photographed, they usually turn the head slightly to the right so as to watch the camera out of the left side of their visual field.

By using an instrument called a tachistoscope to flash pictures

---

3. An object on the left side of your body (that is, in your left visual field) is projected on to the right side of the retinal field at the back of the eyeball. When an image passes through the lens of the eye, it is inverted: we actually see the world reversed and upside-down.

of actors' faces on to particular parts of the retina, Jim Denman and John Manning have shown that people are much more likely to identify correctly the emotion being expressed by the actor when the picture is flashed on to the right half of the retina than on to the left half (or vice versa for left-handed people). In another study, one of my students, Catherine Lowe, has shown that mothers carrying young babies are much more likely to detect a soundless grimace by the baby if it is being cradled on the left side of her body than on the right. (This explains why most people, and mothers in particular, cradle their babies on the left side of the body. The alternative suggestion – that it is to comfort the baby by allowing it to hear the mother's heartbeat and so remind it of its time in the womb – cannot be right: the heart lies in the centre of the chest, not on the left side as is often popularly assumed).

This asymmetry in emotional cueing is clearly of very ancient origin because it is already present in other monkeys and apes. Mark Hauser has shown that the left side of the face tends to respond sooner and more intensely than the right side when monkeys give facial signals such as grimaces. The left side of the face is, of course, under the control of the right hemisphere of the brain. This asymmetry in sensitivity thus long predates the appearance of the hominids, never mind language-toting humans; it may even predate the origin of the primates. All this suggests that the right side of the brain was already heavily preoccupied with the business of monitoring and controlling emotional responses. Since it is the emotional behaviour of another animal that tells us what its intentions are, this is hardly surprising: being able to read these signals correctly, and then to respond to them with an equally emotion-driven response, is what primate social life is all about. Herein lie the beginnings of the hierarchy of intensionality we met in Chapter 5.

Given this, it seems natural that language should localize in the left hemisphere. There was simply more free space there for setting up the specialized neural control centres it required. I suggest that the reason we are so predominantly right-handed is because language allowed a special kind of thoughtful consciousness to develop in the left hemisphere. This in turn made it possible to exert greater control over right-handed throwing than left-handed throwing, so favouring the right side.

The psychologist Julian Jaynes argued along somewhat similar lines in his seminal book *The Origin of Consciousness in the Breakdown of the Bicameral Mind*. For example, he used literary evidence from the Middle East to suggest that the person who wrote the Homeric poems of ancient Greece in around 1200 BC was not fully conscious. Jaynes cites the striking lack of introspection in all the writings from this period: they do not refer to emotions, but instead give straightforward narrative descriptions. Consciousness, he suggested, developed around the beginning of the first millennium BC as the left (linguistic) hemisphere gradually succeeded in exerting control over the more unruly (emotional) right hemisphere. My own feeling is that Jaynes is on the right lines, but that his chronology confounds two separate events: the growing dominance of the conscious left hemisphere (which must have occurred when language evolved) and people's abilities to give expression to their inner emotional states.

This contrast between the control centres for language and emotional behaviour has an interesting and rather unexpected consequence. It turns out that although language is localized on the left side of the brain, music (and poetry) is localized on the right side. In a neat study carried out some years ago, Thomas Bever and Robert Chiarello showed that untrained musicians (people who have had less than three years of music lessons during their lives) recognize tunes more quickly if they are played to the left ear through a set of earphones (hence monitored by the right hemisphere of the brain) than to the right ear. Trained musicians did not show this to quite such a marked extent, as might be expected from the fact that they had been taught consciously to analyse musical excerpts into their component parts rather than hearing the tune as a unified whole.

This suggests that while the conscious manipulation of music goes on in the left hemisphere (along with speech and other 'conscious' activities), the emotional response to the tunefulness and rhythm of music goes on in the right hemisphere. Similarly, there is evidence to suggest that the music of poetry is handled in the right hemisphere, whereas the linguistic content – the words – is dealt with on the left side. This is why, when stroke patients lose their speech following a left-sided stroke, nursery rhymes are used to encourage them to speak again; these simple forms of poetry and song are more likely to be stored in the undamaged right hemisphere.

Moreover, the fact that music is located in the right hemisphere is one good reason why the alternative suggestion that language evolved from song cannot be wholly right. Words are certainly used in song, but words are dealt with in the left hemisphere. Song (and music generally) certainly arouses our emotions in a way that words alone find it difficult to do; and song can be used in a very powerful way to express a group's collective emotional arousal. But it's hard to see how something localized in the right hemisphere can produce something localized in the left hemisphere. A more plausible suggestion is that when language evolved it was hijacked by the emotionally powerful music centres to produce song, because music and song are a very potent means of expressing a group's emotional response.

All in all, gestural theories of language origin seem pretty implausible. In any case, we can point to precursors for almost all the features of human verbal communication in the vocalizations of Old World monkeys and apes. This makes vocal theories of language evolution more plausible right from the outset.

Recall the vervet monkeys that we met earlier. Dorothy Cheney and Robert Seyfarth have been able to show that vervet vocalizations do convey meaning. They are not simply expressions of emotional state. The alarm calls refer to specific types of predator, and the hearer knows from the auditory information alone which type of predator the caller is describing; contact grunts specify a great deal about the ongoing situation.

To this we can add the conversational patterns of the gelada. Everyone who has ever seen gelada has commented on the extraordinary calling sessions they engage in. Bruce Richman has shown that the calls in these sequences are timed very closely to fit between the calls of others. Richman concluded that the animals must be timing their calls by anticipation rather than simply by response to the previous caller. This is one of the characteristic features of human conversation: two speakers intersperse their segments of speech and their interjected comments ('Oh!', 'Well, I never!') so that (most of the time) only one person is talking. The flow of conversation is almost continuous. They achieve this by anticipating the end of the other speaker's phrase or sentence, often using cues that the speaker provides. These include a slight rise in pitch towards the end of a sentence, and the tendency for the speaker to glance at the listener a few words short of the point when he or she is ready to stop talking.

Another bastion of the uniqueness of human language is the production of vowel sounds. Without them, language would be impossible; vowels allow us to divide up the flow of sounds into easily discriminated segments (syllables) and hence to form words. It has long been held that monkeys and apes cannot make these sounds because their mouths and throat cavities are the wrong shape. However, several recent studies of the way monkeys such as the gelada and the macaques produce sounds have shown that, in fact, they do make these vowel-like sounds. This implies that the machinery to produce the sounds of human speech was in place long before humans were even a twinkle in the eye of evolution. The problem is not the mechanics of producing the sounds so much as the machinery for co-ordinating sound production and the cognitive mechanisms for attaching meanings to sounds of this kind (although given the vervet studies, this last point must remain in some doubt).

In other words, we can already see many hallmarks of human speech in the Old World monkeys. In the vervet's calls we have an archetypal proto-language. Quite arbitrary sounds are used to refer to specific objects, to convey information about who is doing what (or about to do what). In addition, these calls can be overlaid with varying degrees of emotional overtone, much as our own verbal statements are. There is no need for a gestural phase. It can all be done by voice. It seems but a small step from here to formalizing sound patterns so that they can carry more information. And from there to producing language is but another small step.

All this suggests that the evolution of the capacity for language was the result of the gradual coming-together of several originally unrelated anatomical and neurological components over a long period of time. No one of them was in itself the trigger for the evolution of language, but each was essential. Had any of them failed to evolve, humans would not be speaking to each other today, and you would not be reading this book.

Ritual and Song

Although we can accept that language began as conventional ape-like vocalizations, there are alternative views about the next step. One is that language evolved as a form of song, reinforcing dance-like rituals that were designed to co-ordinate the emotional states of all the members of the group. The other is that it evolved to exchange information about other individuals in the group. I have so far tended to assume that the second view is the whole story: that language evolved to exchange social information. But it is clear that song plays an especially important role in our lives, so we should consider the merits of the idea.

One of the more intriguing features of human behaviour is the extent to which song and dance feature in our social life. No known society lacks these two phenomena. But when you stop to think about them, they are both very odd activities. Singing is of course something we associate with many animal species, including birds and primates such as the gibbon, whose morning vocal displays we happily refer to as song. For the most part, however, the 'songs' of birds and primates are mechanisms either for defending territory or for advertising for a mate.

To be sure, human song is used for these purpose too. Maasai warriors sing and dance to display their personal prowess before the assembled maidens. And the songs of males in these contexts invariably extol the manly virtues of the singers. We do not often advertise our ownership of land by song, but it is not uncommon for men to sing when going into battle. The Maoris of New Zealand, like many of the Polynesian peoples, used ritual song and dance as a means of intimidating their enemies before battle, a tradition even now maintained by the All Blacks, New Zealand's national rugby team. Scottish regiments were always preceded by their pipers as they marched into battle, a tradition observed as recently as the 1944 D-Day landings in Normandy. We sing our national and club anthems on the sports field with an intensity that we reserve for almost nothing else.

But not all song and dance has this kind of function in human society. We sing in church and around the camp-fire, we sing in bars and in theatres, under circumstances that have little to do with nationalism, battles or the mating game. So why do we do it? Paradoxically perhaps, in answering this question we will find an explanation for the one feature of human behaviour that is *really* difficult to explain: our extraordinary willingness to subject ourselves to someone else's will. The crowd effect is at the same time the most bizarre and the most frightening aspect of human behaviour.

Back in the 1960s psychologists identified a phenomenon that became known as 'risky shift': if you ask someone to express an opinion or do something that is slightly extreme (such as support capital punishment), they will typically express quite moderate views. But if you let them discuss it as a group first, the result will invariably be a much more extreme opinion. We see the same effect in religion: left to their own devices, people will be moderate and tolerant, but in groups their attitudes towards deviance or those of different views become more extreme. The result as often as not is a *jihad*, a holy war against the infidels. To this extraordinary phenomenon we owe the Crusades, Northern Ireland, Rwanda, Yugoslavia and the host of ethnic wars, racial feuds and nationalist vendettas that have sullied the history of our species since time immemorial, not to mention the bizarre business of the *fatwa* against the writer Salman Rushdie.

In an indirect way, the explanation lies, I think, in a little-noticed aspect of song and dance: they are both very expensive activities to perform. Of course, we are all intuitively aware of this. How often have we staggered off the dance floor unable to perform another step? Nor are the singers and musicians spared; they drip with sweat at the end of the performance. Singing, as every aspiring opera diva knows, is hard work: to do it well requires great control of breathing and articulation, and that needs considerable practice. It is, I suspect, no accident that the most evocative and stirring singing is often in the lowest registers of the voice: for example, the bass chanting of the Orthodox Christian liturgy. Deep sounds are difficult to produce, and typically require large body-spaces to serve as resonators. There are well-studied instances in the animal kingdom of deep sounds being used to frighten off opponents by signalling that the caller is a big animal; they involve species as different as toads and red deer. 'Deep croak', as the phenomenon is known to biologists, works because it is a difficult signal to forge: only the largest animals can successfully produce the deepest sounds and sustain the energetic costs of doing so. A deep voice indicates a large and powerful body.

Even we humans give way without a moment's hesitation to those we believe to be larger. We give way to bigger people on the street while barging in front of smaller people. This is a particular problem for women, because of their smaller size; some have mistakenly attributed it to men's 'natural' aggressiveness towards women – but big men do it just as much to little men as to women, and big women are just as likely to do it to little women.

These unwritten rules of behaviour make curious and unexpected appearances in our lives: for example, in no US presidential election since the war has the shorter candidate won. In fact, when Dukakis stood against George Bush in 1988, his managers insisted that he and Bush should not speak at the same lectern when they held a televised debate; a lectern set at normal chest height for the six-foot-something Bush would have practically come up to Dukakis's chin, and the managers were afraid that this alone would lose him the election. Instead, the lectern was lowered after each candidate had spoken, then raised again to the correct height for the next speaker, so that it always appeared at the same relative height on each candidate. Campaign teams are now very careful to ensure that candidates never have to stand side by side on these occasions.

We are often surprised by the shortness of our idols when we actually meet them. The most usual comment people make after meeting the Queen is, 'Isn't she small! I'd expected her to be much taller!' We expect the successful and the powerful to be tall. When asked to choose adjectives to describe successful and unsuccessful people, or leaders and followers, people are much more likely to consider words like 'tall', 'intelligent' and 'attractive' as being true of successful people. And as it turns out, they are not entirely wrong. Research has shown that successful people do tend to be taller, on average. A study in Germany by A. Schumacher showed that, when weighted for social class and educational achievement, senior nurses are taller than junior nurses, skilled carpenters taller than unskilled ones, successful lawyers taller than less successful ones, and senior managers in industry taller than middle managers. Whether this is because taller people are more intelligent or whether it is because we are more likely to give way to them remains to be decided. The fact is that tall people seem to have to work less hard to be successful. If you're shorter than the average, you often have to be bloody-minded to achieve the same level of success – the Napoleon syndrome.

Resorting to deep-bellied sounds is almost universal across human cultures in situations where we want to create a lasting impression. Maori and Maasai warriors hum and grunt in the lowest registers in their war-songs. Successful public speakers do not squeak and warble in falsetto, but lower the pitch of their voice. Adolf Hitler growled his way through his extraordinarily rousing speeches. And it was not for nothing that Margaret Thatcher was trained by her image-makers to lower her speaking voice by nearly half an octave below its natural pitch when she became leader of the British Conservative Party in 1975. Had she not done as they asked, it is quite conceivable that she would not have won such a landslide victory in 1979.

In effect, Mrs Thatcher's minders wanted her to sound more like a man. And this should remind us of the fact that boys' voices 'break' at puberty to produce a deeper, richer tone than women's voices. Quite why this should happen has always been something of a mystery. After all, boys and girls do well enough with their lighter treble voices, and women seem to cope quite adequately throughout their adult lives. 'Deep croak' provides us with a possible answer: human males have been under intense selection pressure to evolve deeper voices during their sexually active years. They have to compete with each other for access to females for mating, and yelling matches have been (and still are) part of the process whereby rivalries are settled. Females have no need to develop deep voices because they do not compete with men or with each other in quite the same way. But they have almost certainly added fuel to the fire of sexual selection by being especially sensitive to deep male voices, thus adding the processes of female choice to the processes of inter-male rivalry (I shall have more to say about sexual selection in Chapter 9).

But there is more to song and dance than simply 'deep croak'. It makes us feel good to sing and dance. It generates euphoric highs, as well as feelings of happiness and warmth. Both activities are hard work; and both are ideal activities to generate surges of opiates from the brain - which is almost certainly why we feel so good after performing them. So why should we have latched on to this curious effect and taken up these activities with so much enthusiasm? The answer, I suggest, lies in the fact that humans live in very large groups, and large groups are difficult to keep together for any length of time. They are perpetually at risk of fragmenting because of the conflicting interests of so many different individuals, not to mention exploitation by free-riders. As the group's size increases, factions with opposing views develop, and we begin to take sides.

Trying to hold together the large groups which the emerging humans needed for their survival must have been a trying business. We still find it difficult even now. Imagine trying to co-ordinate the lives of 150 people a quarter of a million years ago out in the woodlands of Africa. Words alone are not enough. No one pays attention to carefully reasoned arguments. It is rousing speeches that get us going, that work us up to the fever pitch where we will take on the world at the drop of a hat, oblivious of the personal costs. Here, song and dance play an important part: they rouse the emotions and stimulate like nothing else the production of opiates to bring about states of elation and euphoria.

The anthropologist Chris Knight has argued that the use of ritual to co-ordinate human groups by synchronizing everyone's emotional states is a very ancient feature of human behaviour, and coincides with the rise of human culture and language. He argues that to co-ordinate behaviour within a group in the form of ritual requires language, and that this may have been the final stimulus for the evolution of language. To be able to say, as southern African Bushmen do, 'Now let's pretend to be eland in order to do the Eland Dance' (a dance used to celebrate a young girl's first menstruation, her badge of membership of the women's group), it is necessary to have language.

Knight is surely right about the way language is used to formalize and manage ritual. But I am less convinced that language evolved specifically to make ritual possible. My own view is that it evolved to facilitate bonding through the exchange of social information, and was later hijacked for use in these semi-religious contexts in such a way as to formalize what may well have been pre-existing practices. The eland dances of the late Pleistocene would not have been formally construed or tied to specific events. Rather, they were more likely to have been informal, spontaneous affairs, more like Saturday afternoons on the terraces of the local football club. These dances seem to me to be very ancient rituals indeed. Language allowed us to formalize their spontaneity, so giving them more coherence by providing them with a metaphysical or religious significance. But this, I think, must have happened during the Upper Palaeolithic Revolution, when we see the first evidence for religious beliefs and symbolic thought.

And this unexpectedly brings us back to one of the more curious aspects of language, namely its complete inadequacy at the emotional level. It is a most wondrous invention for conveying bald information, but fails most of us totally when we want to express the deepest reaches of our innermost souls. We are so often 'lost for words' in such circumstances. Language is a wonderful introduction to a prospective relationship: we can find out a great deal about a person with whom we might be thinking of forming a relationship or an alliance. But when the relationship reaches the point of greatest intensity, we abandon language and return to the age-old rituals of mutual mauling and direct stimulation. At this crucial point in our lives, grooming – of all the things we inherit from our primate ancestry – resurfaces as the way we reinforce our bonds. We use it because physical contact is deeply moving and reassuring in a way that language cannot be. And it achieves that because monotonous stroking and rubbing stimulate opiate production much more effectively than words can ever do.

Paradoxically, it seems that just as we were reaping all the advantages of language, we had to back-pedal on abandoning the more primitive processes. Just as we were acquiring the ability to argue and rationalize, we needed a more primitive emotional mechanism to bond our large groups and make them effective. Language allowed us to find out about each other, to ask and answer questions about who was doing what with whom. But of itself, it does not bond groups together. Something deeper and more emotional was needed to overpower the cold logic of verbal arguments. It seems that we needed music and physical touch to do that. We have the most remarkable computing machines in the natural world, the most articulate communication system, the most sophisticated minds, yet we are ultimately dependent on crude hormonal tricks to ensure that our groups remain coherent and focused on the common goal that we so urgently need in order to survive and reproduce effectively.

Maiden Speech

All this raises an interesting question: if language evolved to facilitate group cohesion, who spoke first? The bison-down-at-the-lake view of language naturally assumes that it was the males trying to co-ordinate their hunting activities. However, in most primate species it is the females that form the core of society, creating the group and giving it coherence through time. The males, by contrast, are less constant in their social affinities, often wandering from group to group in search of better mating opportunities. In most species (chimpanzees appear to be one of a handful of exceptions), females remain in the groups into which they were born, whereas males typically leave at puberty and transfer to another group. Some males even continue to move from group to group throughout most of their lives.

If females formed the core of these earliest human groups, and language evolved to bond these groups, it naturally follows that the early human females were the first to speak. This reinforces the suggestion that language was first used to create a sense of emotional solidarity between allies. Chris Knight has argued a passionate case for the idea that language first evolved to allow the females in these early groups to band together to force males to invest in them and their offspring, principally by hunting for meat. This would be consistent with the fact that, among modern humans, women are generally better at verbal skills than men, as well as being more skilful in the social domain.

However, the current consensus amongst evolutionary anthropologists is that humans do not have female-bonded societies in this sense. The evidence for this is that patrilocality (brides moving to the husband's village) is characteristic of most, even if not all, traditional societies. However, this may be a consequence of the fact that most of these cultures live in conditions where males are able to control the resources (land, hunting grounds) that women need for successful reproduction. Several lines of evidence suggest that in more equitable economies (such as those of hunter-gatherers and modern industrial societies) female kinship and alliances are much stronger, and males may as often as not be forced to move to their wives' villages.

Several lines of evidence support this suggestion. Among central African pygmies, for example, Y-chromosome genes are more widely distributed than X-chromosome genes, which are much more clustered in their distribution. This suggests that the women have tended to remain close to their female kin groups, whereas men have moved across wide areas in order to mate (so distributing the genes on their Y-chromosomes much more widely). A similar finding emerges from sociological studies carried out in the slums of east London during the 1950s: not only were close female kin (mothers, daughters, sisters, aunts, nieces) crucially important in enabling a woman to live and reproduce successfully in this impoverished environment, but married women lived significantly closer to their own parental homes than to their husband's parents.

We have detected elements of this in our own research. In a study of social networks, Matt Spoors and I found that women not only had slightly larger networks of regular contacts (people they contacted at least once a month) than men, but these also included a higher proportion of their immediate same-sex kin (the limit being set at cousins). Even more impressive evidence comes from an experimental study we carried out in collaboration with Henry Plotkin, Jean-Marie Richards and George Fieldman (the details of which are given on page 164). We found that women were prepared to incur almost as much physical pain to provide a financial reward for a female best friend as for themselves, but men were not nearly so altruistic towards their male best friend.

Taken together, these findings suggest that female bonding may have been a more powerful force in human evolution than is sometimes supposed. If so, then the pressure to evolve language may well have come through the need to form and service female alliances, as Chris Knight suggests, rather than through either male bonding or male hunting activities, as conventional wisdom has always assumed.

It still remains unclear whether the initial stimulus to language was provided by the emotional uplift of the Greek chorus or by the need to exchange information about other members of the alliance or group. The fact that monkeys like the gelada commonly use contact calls in choruses that often rise to extraordinary heights of emotional crescendo might be taken as some evidence to favour the song hypothesis. When gelada reproductive units contact call, usually only one or two animals are involved. But every now and then, the whole group will come together in a rousing chorus of grunts and moans, capped at the end by the harem male's dramatic vibrato yawn which acts like a full stop at the end of a collective sentence. But it is the females, counter-calling to their grooming partners within the unit, that start the process off and for whom it seems to be most important.

This move from simple contact calling of the kind seen in savannah baboons to the emotionally charged choruses of the gelada would seem to be an equally natural development for our own ancestors when grooming time requirements slid inexorably beyond the time actually available as group sizes increased through the Pleistocene (see page 114). If so, then it seems that it was the females who provided the real impetus along this route.

Although male monkeys and apes commonly form coalitions, they seem to be much less tightly bonded than those of females. Frans de Waal's studies of the Arnhem Zoo colony of chimpanzees show this rather clearly. He found that females' alliances were more long term and based on kinship, whereas males' alliances were shorter-lived and based on expediency and the demands of the moment. Alliances of the latter kind require only that the parties recognize the situation and what their ally is really trying to do. Long-term alliances, on the other hand, depend on deeper bonding processes: that almost certainly means a more emotional basis.

But at some point this purely emotional behaviour must have given way to true language and the exchange of information. It is difficult to see how the very large groups that began to emerge towards the end of the Pleistocene could have been bonded successfully without the ability to exchange social information. Important as emotional bonding is to social meshing in our small inner circle of intimate friends, there appears to be a limit on the number of people we can bond with in this way (the so-called 'sympathy-group' of 10 to 15: see page 76). Managing our relationships with the outer circle of acquaintances depends much more heavily on social knowledge than on emotional empathy.

CHAPTER 8

Babel's Legacy

The oddest feature of human language is its habit of producing mutually incomprehensible tongues with astonishing speed. Barely two millennia separate French and Italian from their mutual ancestor Latin, yet most native speakers of these two closely related languages could no more understand each other than they could Latin. Danish and Swedish descend from different local dialects of Scandinavian, yet after barely a thousand years they are now all but mutually incomprehensible. Read Chaucer's *Canterbury Tales* in the original, and you will be only too aware of the extent to which English has changed in 600 years; half the words are unrecognizable. Even Shakespeare, at a mere 400 years' remove, can prove disconcertingly opaque at times.

Our task in this chapter is to try to account for this phenomenon.

The Descent into Babel

The mythologies of many peoples around the world share a belief in a common origin for mankind. Most of these stories imply (though they seldom say so explicitly) that at some remote point in history everyone spoke a common language. The biblical story of the Tower of Babel actually makes a point of telling us that all humans once spoke the same language. Chapter XI of the book of Genesis gives the story:

> And the whole earth was of one language, and of one speech. And it came to pass, as they journeyed from the east, that they found a plain in the land of Shinar; and they dwelt there ... And they said, Go to, let us build us a city and a tower whose top may reach unto heaven; and let us make us a name, lest we be

scattered abroad upon the face of the whole earth. And the Lord came down to see the city and the tower, which the children of men builded. And the Lord said, Behold, the people is one, and they have all one language; and this they begin to do: and now nothing will be restrained from them, which they have imagined to do. Go to, let us go down, and there confound their language, that they may not understand one another's speech. So the Lord scattered them abroad from thence upon the face of all the earth; and they left off to build the city. Therefore is the name of it called Babel; because the Lord did there confound the language of all the earth.

The Tower of Babel was no myth: it really did exist. Its real name was Etemenanki (meaning 'the temple of the platform between heaven and earth'), and it was built some time in the sixth or seventh century BC during the second great flowering of Babylonian power. It was a seven-stage ziggurat, or stepped pyramid, topped by a brilliant blue-glazed temple dedicated to the god Marduk, by then the most powerful of the local Assyrian pantheon. A century or so later, in about 450 BC, the Greek historian Herodotus struggled up the steep stairways and ramps in the hope of seeing an idol at the top. Alas, there was nothing but an empty throne.

However, the myth-makers of ancient Israel seem to have been on to something. Linguists now believe that the world's languages do in fact have a common origin, though it is only recently that they have been able to do more than speculate about the history of language. However, the period of this common language long predates the building of the Tower of Babel, and it is difficult to see why real events in sixth-century BC Babylon should have become associated in the minds of the later writers of the book of Genesis with what might be genuine folk memories of a time when everyone spoke a common tongue.[1] By the time when the Babylonians built the tower of Babel most of the world's

---

1. That tribal folk memories can retain knowledge of events that happened in the distant past is well illustrated by the fact that, according to the archaeologist Josephine Flood, the origin stories of some Australian Aboriginal tribes contain surprisingly accurate descriptions of the land surfaces below the Tasman Sea off

major language groups were already well established.

The reconstruction of the history of languages dates back to Sir William Jones, a scholarly and enquiring man who was appointed a judge in Calcutta towards the end of the eighteenth century. Convinced that he ought to study Hindu legal authorities in the original if he was to give proper legal judgements, he set about learning Sanskrit, the ancient (but by then dead) language of northern India. As his learning progressed, Jones became increasingly struck by the similarities between Sanskrit and the ancient European languages, Greek and Latin, that he knew so well. Making due allowance for sound changes, he was able to identify sufficient similarities between the words of each of these languages to claim that they derived from a common mother tongue.

Classic examples include words like 'brother' – which has clear similarities to the Greek *phrater*, the Latin *frater*, the Old Irish *brathir*, the Old Slavic *bratre* and the Sanskrit *bhrater* – and the parts of the verb 'to be', which include *is* in English and Old Irish, *esti* in Greek, *est* in Latin, *yeste* in Old Slav and *asti* in Sanskrit. In most cases, the slightly different forms of these common words are the result of consistent shifts in pronunciation, a point first noticed by the folklorist brothers Jakob and Wilhelm Grimm. Examples of these shifts were the Latin *f* and *ph* sound changing into the *b* of English and other Germanic languages, and the *p* and *t* sounds of the ancestral languages (as in Sanskrit *piter*, Latin *pater*) changing to the *f* (or *v*) and *th* sounds of the later Germanic languages (German *Vater*, English *father*).

The scholars of the nineteenth century were inspired by this suggestion, and spent much time trying to reconstruct common

the coast of southern Australia, as well as accounts of how the seas rushed in to sever the land links between the mainland and many of the islands off the northern and southern coasts. During previous ice ages, the sea bed in these areas would have been exposed as dry land. The last time the Aboriginals' ancestors could have walked on them was around 10,000 years ago, just before they sank beneath the waves for the last time as the ocean levels rose with the melting of the ice-sheets at the close of the last Ice Age. Moreover, it seems that the Australian Aboriginals are not alone in this respect. The Scandinavian *Ragnarök* legend gives an account of a time known as the *fimbulvinter* when severe winters followed each other in succession without intervening summers; it has been suggested that this is a folk memory of the 'little ice age' that afflicted northern Europe around 1000 BC.

ancestors for various language groups. So much so, in fact, that by the end of the century the academicians of the Société de Linguistique in Paris were exasperated enough to ban speculations about the history of languages from their meetings. But by the time historical linguistics fell out of fashion, it had been possible to reconstruct the relationships between most of the world's better-known European and Asian languages. It was soon recognized, for instance, that the languages of most of Europe (except Basque and the Uralic-Yukaghir language group that includes Hungarian, Finnish and Estonian) and those of southern Asia as far east as the Indian plains (including Persian and the modern derivatives of Sanskrit) all belonged to the same group, now known as Indo-European.

This ancient language is thought to have originated somewhere north of the Danube basin in about 5000 to 6000 BC. Indo-European languages share the same words for winter, horse and domestic animals such as sheep, pigs and cattle, as well as words associated with leatherworking, ploughing and planting grain. Taken together, these suggest that the ancestral Indo-Europeans led a semi-nomadic existence in which agriculture played a prominent role. Their gods were clearly the ancestors of the Indian, Mediterranean and Celtic deities; and they had a well-established social structure, since their descendant languages share many of the words for kinship and family structure. One thing they do lack, however, is a common word for 'sea': it's clear that the Indo-Europeans did not live near any major ocean or lake coasts.

During the last few decades, there has been a renewed interest in reconstructing language trees. The consensus now seems to be that Indo-European and the other language groups of Europe (Basque, Uralic, etc.) and Asia (the Semitic languages of North Africa and the Near East, the Altaic languages that include Turkic and Mongolian, and the Elemo-Dravidian languages of southern India) represent the descendants of a superfamily known as Nostratic which probably originated around 13,000 BC.

Thanks to the hard work of a group of Russian linguists, attempts to reconstruct Nostratic have been quite successful, at least in the sense of being able to produce a vocabulary of several thousand words that appear to be plausible common ancestors for the appropriate words in the modern Eurasian languages of this family. These include words such as *tik* meaning finger (or one), from which the modern English word *digit*, the French *doigt* and the Latin *digitus*, as well as the Hindi *ek'* (one), are all said to derive. The proto-Indo-European *melg*, 'to milk', has similarities with the proto-Uralic *malge*, 'breast', and the modern Arabic *mlg*, 'to suckle', again suggesting a common origin. Of particular interest is the fact that these languages share words for certain kinds of things but not others. There are words for dog – *kujna,* from which derive German *Hund* and English *hound* – and for things connected with the outdoors, such as *marja*, 'berry', but there appear to be no words associated with farming. This suggests that the people who spoke Nostratic were hunter-gatherers who lived in the period prior to the discovery of agriculture around 10,000 years ago.

In addition to Nostratic, four other non-African language superfamilies are now recognized: Dene-Caucasian (the languages of the Arctic and sub-Arctic regions of Asia and North America), Amerind (most of the native languages of the Americas south of the Canadian border), proto-Australian (the languages of the Australian continent) and Austro-Asiatic (the languages of southeast Asia). The languages of Africa south of the Sahara are more confusing, and their position with respect to these five superfamilies remains unclear. But three main families have been identified: Khoisan (the languages of the southern African Bushman and related peoples), Niger-Kordofan (the Bantu languages that dominate most of western, central and eastern Africa) and Nilo-Saharan (the languages of the largely nomadic peoples on the southern rim of the Sahara desert).

There have even been attempts to reconstruct the common original of these eight superfamilies, the so-called 'proto-World' tongue. Linguists like the Russian Vitaly Shevoroshkin, now at the University of Michigan, claim to have been able to identify some 200 words of proto-World. These include *nigi* (or *gini*), 'tooth', with its derivates in the Congo-Saharan *nigi*, the Austro-Asiatic *gini*, the Sino-Caucasian *gin* and the Nostratic *nigi* (from which may derive the modern English words *nag* and *gnaw*). Similarly, the English word *tell* has been related to the proto-World *tal* (or *dal*), 'tongue' – telling being an important activity for your *tal*.

The reconstruction of long-dead languages is, of course, a chancy business, and attempts such as those I've described have not been without their critics. Indeed, they have often been condemned as fanciful. Some linguists point out that the rate at which languages change naturally over time is such that after about 6000 years it would be virtually impossible to identify any common terms. Whether we will ever be able to reconstruct the original languages is perhaps less important than the fact that we can trace common ancestries for many of them. Although it may be fun to converse in Nostratic (as the Russian linguists regularly did), the more significant question is *why* languages should have diversified to produce so many thousands of mutually incomprehensible tongues from a single common origin.

The Dynamics of Chaos

There are estimated to be about 5000 languages currently spoken in the world today, depending on which you count as dialects and which as fully fledged languages. To these, you can perhaps add a handful of 'dead' languages that are still taught in schools (ancient Greek and Latin) or used in religious services (Sanskrit and Ge'ez). Linguists reckon that well over half of all these languages will become extinct, in the sense of having no native speakers, within the next half-century. They are mostly languages which currently have fewer than a thousand native speakers, most of whom are already elderly. The time may come, it has even been suggested, when the world will be dominated by just two languages; on present performance, these will almost certainly be English and Chinese. The loss of all these languages will, of course, be a pity. As we lose them, we lose fragments of our past, for languages represent the history of peoples, the accumulation of their experiences, their migrations and the invasions they have suffered.

But this observation overlooks one curious feature of human behaviour: our tendency to generate new dialects as fast as we lose others. English has spread around the globe to become the *lingua franca* of trade, government and science, as well as the national language of countries on every continent; yet, at the same time, it has diversified to produce local dialects that border on the mutually incomprehensible. Most linguists now recognize Pisin (the 'pidgin English' of New Guinea), Black English Vernacular (BEV, the form of English spoken by blacks in the major cities of the US), Caribbean creoles (the English of the various Caribbean islands) and Krio (the creole of Sierra Leone in West Africa) and even Scots (the English spoken in the Scottish lowlands) as distinct languages. A thousand years from now, some historical linguist may trace their origins back to an obscure island off the north-west coast of Europe, and wonder how it came to be that the language of one unimportant member of the vast Indo-European language family came to oust all the others.

In fact, this process of dialectization is not unique to human languages. We now know that the 'languages' of other species also have dialects. Crows in eastern Europe pronounce their caws in a noticeably different way from those in western Europe, while the Japanese macaques in the northern parts of their range pronounce their *coo* contact calls differently from those in the south. Of course, the scale of this is quite limited by comparison with the extent and rapidity of the dialect evolution in human languages, but the pattern is closely analogous. This flexibility is so striking and so universal that it cannot be a simple accident of evolution: it must have a purpose. A clue to this may lie in the history of language evolution itself.

The archaeologist Colin Renfrew has argued that the modern language groups evolved as a result of four major migration events in human history. The language groups of Australia and the Amerind languages of the New World owe their origins to the initial migration that saw modern humans spilling out of Africa around 100,000 years ago. The leading edge of this dispersal moved gradually eastwards until about 40,000 years ago it more or less simultaneously slipped across the Bering Strait into North America and across the Arafura Sea that separates Australia from the islands of the Sunda Shelf which juts out below the Indo-Chinese peninsula. Other remnants of that same early dispersal, he suggests, include Khoisan (the language of the Southern African Bushman and their relatives), Basque (arguably descendants of the original inhabitants of Europe – heaven help us if they demand their rightful tribal lands back!), Caucasian (the languages spoken in the region between the Caspian and Black Seas), the Indo-Pacific languages of New Guinea, and the Austric language group (the ancestral languages still spoken by hill tribes in parts of Vietnam, Cambodia and Thailand).

The second great event occurred about 10,000 years ago, following the more or less simultaneous development of farming at various sites in the New and Old Worlds. The discovery that certain plants could be farmed at will freed the peoples concerned from the inexorable need to follow the migrating game and the fruiting cycles of the woodland trees. Farming allowed groups to stay longer in the same place; moreover, the surpluses produced by farming allowed their populations to grow faster. As these peoples spread outwards from the regions of initial settlement in search of new land for their burgeoning families, they displaced (or probably less commonly absorbed) the hunter-gatherer communities they encountered. We have ourselves witnessed analogous cases of physical displacement in recent historical times: the migrations of the Saxon peoples into western Europe during the fifth and sixth centuries AD, when they gradually pushed the Celts into the peripheral hill regions in the north and west of the British Isles, and the nineteenth-century European migrations to North America and Australia.

The most important of these movements were associated with grain farming in the Near East – which gave rise to the Afro-Asiatic language group that spread south-westwards into Arabia and North Africa, and to the Indo-European group that spread west into Europe – and with rice farming in the Far East, which produced the Sino-Tibetan languages of southern China and, indirectly, the Austronesian languages of the Pacific rim, including offshoots like the languages spoken as far apart as New Zealand and Madagascar.[2]

2. Despite being off the south-east coast of Africa, the island of Madagascar was in fact first settled by people from the Pacific rim about 2000 years ago. Their language, Malagasy, is closely related to some of those spoken today in Borneo, and they probably reached the then-uninhabited Madagascar as a result of early trading expeditions.

The third series of migrations occurred around 8000 years ago. The global warming that spelled the end of the last Ice Age opened up the Arctic regions to inward migration by a group of people who proceeded to disperse westwards into northern Scandinavia, where they are now represented by the reindeer-herding Lapps, and eastwards across the Bering Strait into the Canadian Arctic, where they displaced the native Amerindians southwards into what is now the USA. Clearly, the latter migration must have taken place in at least two major waves: an earlier group represented by the Na-Dene language group of (mainly) Canadian Indian peoples, and considerably later the Eskimo (or more correctly Innuit) peoples who now occupy the Arctic regions from Alaska to Greenland.

Finally, within historical times a number of major migrations gave rise to what Renfrew terms the 'élite dominance' process. The development of complex societies, often associated with the use of the horse as a means of travel and warfare, led to rapid movements of Indo-European peoples eastwards into the Middle East and northern India, and of the Altaic-speaking peoples of Mongolia eastwards from central Asia into northern China, Siberia and the Japanese archipelago. This latter group were responsible for a second major expansion, this time westwards as far as eastern Europe, during the twelfth century AD as a result of the empire-building activities of the much-feared Mongol leader Genghis Khan. In most of these instances, the invaders enslaved or absorbed the native populations of the regions they took over rather than driving them out. As a result, the invaders imposed their language on their vanquished subjects (much as the Spanish and Portuguese imposed their languages on the Indians of South and Central America).

What this seems to suggest is that, at least until recently, it has been the wholesale migration of peoples that has been largely responsible for the spread and diversification of languages. Elite dominance is a recent phenomenon, reflecting developments in transport and technology; the wholesale replacement of one group of people by another, bringing their own language and culture with them, seems more often to have been the norm. This suggestion finds dramatic support from an unexpected direction. During the late 1980s the ability to sequence the base pairs on segments of DNA made it possible to carry out detailed comparisons of the genetic codes of different species. In a startling breakthrough, Rachel Cann, the late John Wilson and their colleagues at Stanford University in California mapped out the sequence of base pairs on a segment of the mitochondrial DNA of 120 or so women who had given birth in local hospitals. The important thing about mitochondria[3] is that they are passed down only through the female line, so a pedigree constructed from their similarities is a direct reflection of the mother-daughter descent lineages involved. By comparing the sequences from different individuals they were able to reconstruct the genetic relationships between those individuals, and hence between the many different racial groups represented among the mothers in the sample.

These analyses revealed that the range of variation in mitochondrial DNA is much greater in Africa than anywhere else in the world. Everyone from Europe, Asia, Australia and the Americas, plus some north African peoples, seems to be part of a single, closely related family group, which itself is a subset of the broad family of African peoples. By working backwards through the hierarchy of relationships, Cann and Wilson arrived at a common female ancestor for the women, who was inevitably named the 'African Eve'. Finally, by determining the number of mutations along each of the descent lines, they were then able to use the natural rate of mutation for mitochondrial genes, the so-called 'molecular clock', (see page 33) to estimate that this common female ancestor probably lived between 150,000 and 200,000 years ago.

Although there has been some dispute about how the reconstructions were done and about the methods used to determine when the ancestral Eve lived, the original hypothesis has largely been vindicated by subsequent analyses based on larger samples of women. More importantly perhaps, it is in good agree-

---

3. Mitochondria are the tiny power-houses within each cell that provide its energy. It is now believed that they were originally viruses that invaded the cells of some very ancient single-celled organisms and stayed on to develop a symbiotic relationship with the cell's own nuclear DNA.

ment with the fossil record: the only fossils that could be ancestral to modern humans come from Africa during the period between 250,000 and 150,000 years ago.

Strictly speaking, these analyses do not identify a single common ancestor: they merely tell us that the mitochondrial DNA of all living humans – males get theirs from their mothers too – derives from a very small number of females living at a particular time. Other calculations suggest that the ancestress(es) of all modern humans were members of a population that totalled only about 5000 individuals of both sexes and all ages. These would not have been the only members of their species (our species) alive at the time, but they were the only ones whose lineage(s) survived to pass on their mitochondrial DNA to us.

The real surprise came when the geneticist Luigi Cavalli-Sforza showed that if you plotted the tree of language groups on top of Cann's tree of the genetic relationships between human racial groups, the fit was surprisingly good. The dispersal and divergence of the main language groups appeared to mirror the dispersal and divergence of the racial groups concerned. This suggests that when people migrated, they took both their genes and their languages with them, replacing wholesale the other human populations that they came across. Thereafter, as their genetic make-up diverged through the accumulation of mutations, so their languages diverged by a process of dialectical drift.

And this brings us back to the question of why human languages develop dialects so readily. The changes in pronunciation over time documented by the brothers Grimm and other historical linguists are one of the mechanisms that have produced the extraordinary diversity of languages we have today. It raises the fundamental question of why dialects evolve.

My Brother and Me

It is widely recognized that dialects are intimately related to local subcultures. Speaking a particular dialect is a badge of group membership. It shows we belong. But why should we be so concerned to establish our group membership? And why should dialects be so effective in this respect? The answer to the first question has to do with the Enquist-Leimar free-rider problem. But the answer to the second question – why dialects are an effective solution to that problem – depends on an important feature of evolutionary biology we have not so far encountered: the theory of kin selection.

Mating and reproduction are obviously important from an evolutionary point of view. They are, after all, the way you pass your various genes on to the next generation. But they are not the only way of doing so. In the mid-1960s the New Zealand entomologist Bill Hamilton (then a young postgraduate student at Imperial College London, but now a professor at Oxford University) pointed out that an individual can ensure that the genes it carries are transmitted to the next generation in either of two ways: by reproducing itself, or by helping a relative who carries the same gene to reproduce more successfully.

Provided the cost of helping a relative to reproduce (measured in terms of the lost opportunities to reproduce that the helper incurs) is less than the gain to the relative (when devalued by the degree of relationship between them), then it will pay an individual to behave in a helpful way. This principle is known to evolutionary biologists as 'Hamilton's Rule': it specifies the conditions under which helping behaviour can be expected to evolve within a population of animals. The mechanism is known as kin selection; it is not the only Darwinian mechanism for the spread of altruistic behaviour, but it does have important implications for the behaviour of organisms in general.

The key consideration in Hamilton's Rule is whether the two individuals share a particular gene. The more closely related they are, the more likely it is that they will have inherited the same gene from a common ancestor, and hence the more worthwhile it is for the one individual to help the other. The main point here is that, all things being equal, it is more beneficial to help a close relative than a distant one, because close relatives are more likely to share a given gene than distant relatives or, worse still, unrelated individuals.

The importance of this for our story is that most higher organisms, including humans, show a strong preference for relatives. In general, humans prefer to live near relatives rather than non-relatives - although there obviously has to be some negotiation over *which* set of relatives a married couple lives near. It is relatively rare in pre-modern societies, for example, for newly married couples to live apart from both sets of relatives. Even in modern post-industrial societies, where market forces mean that many people have to move away from their parental homes to find jobs, significant numbers continue to live near their relatives if they can, and those who have moved maintain contact with relatives for longer than they do with unrelated friends. Moreover, people are decidedly more likely to help relatives than less closely related individuals. 'Blood is thicker than water' is a sentiment repeated in almost every human culture. As the Arab proverb so graphically reminds us, 'Me against my brother; me and my brother against my cousin; me and my brother and my cousin against [our common enemy]'.

There is a great deal of evidence to show that humans take relatedness into account in their dealings with other humans, especially when the costs of behaving altruistically are high. This is not to say that humans are never altruistic towards other humans. They quite often are, but normally only when the costs of doing so are minimal. People like Sydney Carton, from Dickens' stirring story of selflessness *A Tale of Two Cities*, are comparatively rare in real life. Were altruistic behaviour towards all and sundry truly 'natural', we would not have to be exhorted to act altruistically as often as we are; such behaviour would simply be taken for granted. And we would all willingly pay our taxes in full and on time.

In order to study Hamilton's Rule at work in humans, Henry Plotkin, Jean-Marie Richards, George Fieldman and I carried out a simple experiment that had originally been suggested some years earlier by the biologist David McFarland. Subjects were asked to perform an isometric skiing exercise that involves sitting against a wall with nothing to support them. The thighs are thus parallel to the floor, while the shins and back are at right angles to them. Initially, it's a pleasant enough position to take up. But the leg muscles are under intense stress, and after about a minute the position becomes increasingly painful. Most people can only hold it for two minutes or so before collapsing on the floor.

We offered subjects 75 pence for every 20 seconds they could maintain the position. The catch was that the money they earned was on most occasions given directly to someone else. The identity of the recipient was chosen at the start of the exercise, and so was known to the subject. Each subject performed the exercise six times, each time with a different recipient nominated for the rewards of his or her efforts. The subjects themselves were always one of those recipients, and a major children's charity was always another. The other four recipients were individuals of specified degrees of relatedness – a parent or sibling (related by a half), an aunt or uncle (related by a quarter), a cousin (related by one eighth), plus a best friend of the same sex (relatedness zero).

The results of the experiment were striking: subjects worked much harder for close relatives (parents, siblings) and themselves than they did for less closely related individuals or the charity.

This is of course a very simple situation. But it does capture the essence of real altruism – bearing a cost to benefit someone else – and is probably comparable to many small acts of altruism in which we engage daily, such as lending small quantities of money or giving time to help others.

In another study, Amanda Clark, Nicola Hurst and I looked at thirteenth-century Viking sagas. The histories of the Viking communities in Iceland and Scotland tell of feuds and vendettas that rumbled on for decade after decade. We found that the Vikings were significantly less willing to murder close relatives than more distant relatives, relative to the numbers of individuals available in each category in the population. Only if the stakes were very high, something like the inheritance of a title or a farm, were they prepared to murder close relatives. The relatives of the victim were also more likely to insist on their right to a revenge murder for a close relative than for a distant relative (where blood money was often accepted) – unless the murderer was a particularly violent man, which obviously raised the risks of attempting a revenge killing. These tendencies are reflected even in modern homicide statistics, as Martin Daly and Margot Wilson have pointed out.

Canadian and US murder statistics show that people are twenty times more likely to murder an unrelated person they live with than a genetic relative.

More starkly still, the Canadian data show that children are sixty times more likely to die before the age of two years at the hands of a step-parent than at the hands of a biological parent. This doesn't mean that every step-parent is an ogre; to put all this into perspective, we are only talking about a murder rate of around 600 per hundred thousand children born. But it does mean that the risks step-children run are considerably higher than those run by children living with both biological parents. Something holds us back from uncontrollably venting our frustrations on other people when the source of those frustrations is genetically related to us.

The influence of kinship emerges in other aspects of social life too. When the Vikings made alliances, for example, those formed with relatives were more likely to be stable than those between unrelated people, and they were more likely to be entered into voluntarily without demanding a price in return. Unrelated allies often demanded a favour or property as a precondition, whereas relatives would willingly lend a boat for an expedition or take part in a revenge killing out of a sense of obligation alone. This willingness to help relatives can even be seen in contemporary populations. Catherine Panter-Brick, of the University of Durham, found that among Nepalese hill-farmers the women would willingly help relatives harvest their fields without obligation, but would expect (and demand) a strict reciprocation of help when it came to helping unrelated members of the community.

This emphasis on living with and helping kin is not, I think, simply a reflection of the desire to help relatives reproduce more effectively, in the way that sterile worker bees selflessly help the queen to produce more and more sisters for them. Rather, it seems to be a consequence of the fact that relatives are more likely to co-operate with you in the formation of alliances, as well as in all the minor daily opportunities to help out, because they have a stake in your ability to reproduce successfully. When you need help, you are more likely to get it from a relative than an unrelated individual.

Our kin seem to be of such overriding importance to us that we use the language of kinship to reinforce a sense of group identity even when the other people concerned are not actually related to us. When we do this, it is invariably because those people are valuable allies in some respect. We often refer to co-religionists as brothers and sisters, for example, particularly when we are members of a beleaguered minority. The New Testament letters of the apostles are full of terms of endearment. 'To Timothy, my dearly beloved son', wrote Paul to his completely unrelated colleague; and he began his letter to the Colossians, 'Paul, an apostle of Jesus Christ by the will of God, and Timotheus our brother.' The apostle John opens his First Epistle with the words, 'My little children, these things I write unto you, that ye sin not.' Christianity in particular is infused with this sense of family membership. The most important prayer in the Christian canon opens with the words 'Our Father, which art in Heaven', while on the more earthly level the honorific title 'Father' is given to priests of the Roman Catholic and Orthodox churches.

Interestingly, we resort to the same tactic when rousing our countrymen to the defence of the realm. Terms of kinship and appeals to the protection of our nearest and dearest are so common in this context we barely notice them: 'The Fatherland needs you!'; 'Come to Mother Russia's defence!'; 'They will rape your daughters and sisters!' In order to persuade others to take part willingly, we go to considerable lengths to persuade them that we have kinship ties, even though those ties are at best nebulous.

One problem humans face with their large, diffuse groups is the ease with which free-riders can cheat the system by claiming to be a relative in order to beg a favour. One solution might be to have your DNA fingerprint stamped on your forehead, much as the character Rimmer in the television sci-fi comedy *Red Dwarf* has an H on his forehead to indicate his status as a hologram. Needless to say, this would be rather difficult to arrange; although it's worth noting that Hindu caste marks are just such a sign – and caste membership is, of course, inherited. One way to be sure that someone you encounter really is a relative is to require them to exhibit a token of group membership, some kind of badge. It is, of course, only too easy to cheat with most badges. To be effective, badges either have to be costly to possess (the equivalent of a very expensive motor car) or difficult to acquire (because they require years of practice from a very early age).

In fact, most animals rely on familiarity. Statistically speaking, most individuals with whom you grow up are likely to be relatives. While there will obviously be mistakes from time to time, errors of recognition will only be serious enough to undermine Hamilton's Rule if they become very frequent. The evolutionary process can tolerate a surprisingly high degree of error because it is concerned not with absolute values but with *relative* benefits.

Dialect is an obvious badge, because language is learned at a critical period early in life. Someone who speaks in the same way as you do, using similar words with the same accent, almost certainly grew up near you, and at least in the context of pre-industrial societies, is likely to be a relative. It's not a 100 per cent guarantee, of course, but it's a lot better than simply guessing.

But dialect has another advantage: it can change relatively quickly, at least on the scale of generations. This makes it possible to keep track of the movement patterns of individuals over time. A group that emigrates will evolve its own accent and style of speech over a generation or so, even though it uses the same words. Consider how different the Australian and British accents are now, despite the fact that most of the emigrants to Australia went there within the last hundred years. The obvious suggestion, then, is that dialects were an adaptation to cope with the problem of free-riders. By constantly developing new speech styles, new ways of saying the same old things, a group ensures that it can easily identify its members. And it does so with a badge that is difficult to cheat because you have to learn it early in life. It is not easy to pick up a group's accent or style of speech without living in the group for a prolonged period of time.

To test this idea, Dan Nettle developed a computer model that allowed different strategies to compete against each other in a 'world' where they could only reproduce once they had acquired a certain level of resources (in effect, eating enough food to do more than just keep body and soul together). Individuals were able to co-operate in order to acquire the resources in question, and often had to do so in order to obtain enough to reproduce. Some individuals were co-operators, others were cheats who obtained help but then refused to pay it back. Some of the co-operators only helped individuals who had similar dialects to themselves. But there was a brand of free-riders who quickly learned to imitate the dialects they came into contact with. The model showed that so long as dialects remained constant through time, the imitator-cheat strategy was a very successful one and thrived at the expense of the others. However, if dialects changed moderately quickly (on the scale of generations), cheating strategies found it almost impossible to gain a foothold in a population of co-operators – provided at least that individuals can remember whom they have played against in the past. A dialect that continues to evolve provides a secure defence against marauding free-riders.

It seems likely, then, that dialects arose as an attempt to control the depredations of those who would exploit people's natural co-operativeness. We know instantly who is one of us and who is not the moment they open their mouths. How many times in history, for example, have the underdogs in a conflict been caught out by their inability to pronounce words the right way. As chapter 12 of the Book of Judges tells us:

> And the Gileadites took the passages of Jordan before the Ephraimites: and it was so, that when those Ephraimites which were escaped said, 'Let me go over'; that the men of Gilead said unto him, 'Art thou an Ephraimite?' If he said, 'Nay'; Then said they unto him, 'Say now Shibboleth': and he said 'Sibboleth' for he could not frame to pronounce it right. Then they took him and slew him at the passages of Jordan: and there fell at that time of the Ephraimites forty and two thousand.

In another famous incident, in the courtyard outside the high priest Caiaphas's house in Jerusalem, the disciple Peter was soon identified as an outsider, a Galilean. In the words of the evangelist Mark:

> And a little after, they that stood by said again to Peter, 'Surely thou art one of them [a follower of Jesus, who had just been

arrested]: for thou art a Galilean, and thy speech agreeth thereto. But he began to curse and to swear, saying, 'I know not this man of whom ye speak.'

Nor are these examples confined to the long-distant past. In the Netherlands at the end of the Second World War, not a few German soldiers trying to escape in the mêlée of fleeing civilians came badly unstuck when challenged by Dutch resistance fighters to pronounce complicated Dutch place-names.

Language, then, is very much a social tool. Not only does it allow us to exchange information relevant to our ability to survive in a complex, constantly changing social world, but it also allows us to mark other individuals as friend or foe. And herein probably lie the origins of the historical language evolution we discussed in Chapter 7. Languages diversify initially as local dialects, but eventually these become mutually incomprehensible because local groups need to maintain their identities in the face of competition from other groups. There is some evidence to suggest that, at least in West Africa, language diversity is higher (that is, there are more different languages per square mile, and individual languages have fewer native speakers) in the high-density populations near the Equator than among the low-density populations further to the north, where the proximity of neighbours is much less of a problem.

If this result is confirmed (Dan Nettle is currently testing it), it suggests that the rate at which dialects evolve is not constant, but is directly related to population density. The higher the density of people, the faster their dialects change. The discovery of agriculture a bare 10,000 years ago marks a watershed in human ecology. It had a dramatic effect on population growth-rates, because it allowed people to live at very much higher densities than they had previously been able to do as mobile hunter-gatherers. That being so, it seems plausible to suggest that dialects might even have a rather recent origin. Prior to the agricultural revolution, people might well have spoken the same dialect over a very wide area, with dialect evolution being a slow process of change by gradual drift. Babel may not, in fact, have been so very long ago.

CHAPTER 9

The Little Rituals of Life

We often underestimate just how much of human language depends on an interpretation of the speaker's intentions. Without theory of mind (ToM) and the higher orders of intensionality (see Chapter 5), we would not be able to make more than the barest sense of what others say. Conversations would be factual and dull; they would have all the warmth and poetry of a conversation with *Startrek*'s Mr Spock. We would not have even the most rudimentary literature; the best we could aspire to would be some rather dull narrative poetry. As it is, we use language daily to try to influence the lives of those around us, ultimately for our own benefit.

And herein lies the great enigma of language: what we use for good we can as easily use for evil. With the benefits of Machiavellianism and deep ToM to aid us, we can use language to outwit and bamboozle, to lay propaganda trails to mislead, or to inveigle and cajole. On the whole, I have shied away from exploring these slightly disreputable features of language, preferring instead to concentrate on the more general benefits that language confers in terms of bonding groups. But the time has now come to explore these aspects of human behaviour in more detail.

The Black Art of Propaganda

Free-riders are, as we have seen, an especially acute problem in the large dispersed groups like those typical of modern humans. Trying to prevent them gaining the upper hand becomes a critical problem if your survival in the rough and tumble of the real world depends on maintaining large cohesive groups. Enquist and

Leimar have suggested that gossip may have evolved as a mechanism for controlling the activities of free-riders. By exchanging information on their activities, humans are able to use language both to gain advanced warning of social cheats and to shame them into conforming to accepted social standards when they do misbehave. This is a powerful mechanism for deterring cheats, and Enquist and Leimar were able to show mathematically that free riders would be less successful in a community of gossiping co-operators. Perhaps language evolved not so much to keep track of your friends and acquaintances as to keep track of free-riders and coerce them into conforming.

There is, in fact, some experimental evidence to support this suggestion. Lida Cosmides, of the University of California, Santa Barbara, has suggested that the human mind contains a special module that is designed to detect people who cheat on social agreements. She used an old psychological test called the Wason Selection Task to demonstrate this. In the original Wason Task, subjects were presented with four cards marked with four symbols – say A, D, 3 and 6. The subject is told that there is also a symbol on the back of each card; in addition, they are told that there is a general rule which states that a vowel card always has an even number on its obverse. Which card or cards should they turn over to check that the rule is true?

The logically correct answer is that they should turn over the A card and the 3 card. The A card must have an even number on its reverse and the 3 card must not have a vowel. About three-quarters of those tested on this problem get it wrong (roughly about the number you would expect if people chose the cards at random). Most people choose either the A card or the A card plus the 6 card. But the rule they were given does not say that an even numbered card has to have a vowel on the other side, merely that vowels cards must have an even number on their obverse. An even-numbered card could have either a consonant or a vowel on its back without breaking the rule.

Cosmides was able to show that if you give subjects the same *logical* problem dressed up as a social contract, they generally get the right answer with no trouble. One of her social problems was the under-age drinking rule. Instead of four cards, subjects are told there are four people sitting round a table; one is sixteen years old, one twenty, one is drinking Coca-Cola and one is drinking beer. The social rule is that only those over eighteen can drink alcohol. Which individuals do you need to check to ensure that this rule is not being violated? The answer is trivially obvious: the sixteen-year-old (because sixteen-year-olds are not allowed to drink alcohol) and the beer drinker (because he or she must be over eighteen). Twenty-year-olds can drink whatever they like and anyone can drink Coke. Nearly everyone gets this version of the problem right, despite the fact that they fail miserably on the abstract version of the same problem.

Cosmides has argued that we have an in-built mental set designed to recognize social contract situations and to detect violators. Without this, human social groups would collapse into just the kind of black hole of self-interest that Enquist and Leimar identified. Since cooperativeness is essential to our survival (indeed, it might be regarded as *the* key human evolutionary strategy), we have to have mechanisms for policing observance of the rules we agree for the collective good (where the collective good really means: what's best for each individual in the long run).

This overriding concern with social cheats forces us to consider the fact that there are several ways in which language might work as a social device. I have tended, so far, to assume that language's social function is broadly the exchange of information on friends and acquaintances. But it may be that, as a device for ensuring the stability of large groups, language may in fact produce its bonding effects in a number of different ways. Allowing you to keep track of what your friends and allies are doing is one possibility. But another clearly lies in the exchange of information about free-riders. A third is that language provides us with a device for influencing what people think about us.

The psychologist Nick Emler, for example, has argued that much of our daily use of language is in fact concerned with reputation management. You can pass on information about yourself in order to influence your listeners' perceptions of you. You can tell them about your likes and dislikes, how you would behave (or how you think you *ought* to behave) in different circumstances, what you believe in and how strongly you believe it, what you disapprove of, and so on. You can be deliberately rude or obsequiously nice; you can insult them or flatter them. It can allow you to sort the sheep from the goats very quickly by driving away those whom you know you would never get on with or encouraging those who might be of interest to stay and further their acquaintance with you. Or, of course, you can engage in black propaganda, sowing the seeds of doubt about enemies in people's minds or praising a slightly dubious friend to the hilt so that he or she gets the job.

The fact that we can identify a number of different advantages to language raises the question of whether any one of them was the key selection pressure for language evolution (with the others simply being the metaphorical icing on the evolutionary cake – useful additional benefits, but not of sufficient magnitude of themselves to have driven language evolution on their own). An unequivocal answer would involve demonstrating that, if the other benefits were eliminated completely, language would survive only if, say, the policing function was left in place. With something as complex as language, it is often difficult to disentangle the various functions it can now serve. However, simply asking whether one function is more common than the others might at least provide us with an indication of the likely answer.

In an attempt to throw some light on the problem, Anna Marriott undertook a more detailed study for me of what people talk about. She discovered that criticism and negative gossip accounted for only about 5 per cent of conversation time, with a similar amount of time devoted to soliciting or giving advice on how to handle social situations. By far the most common topics of conversation were who-was-doing-what-with-whom and personal social experiences. About half of this was concerned with other people's doings and about half with the activities of the speaker or immediate audience. This suggests that, whatever else may be going on, monitoring the activities of free-riders and social cheats may not be the primary use to which we put our linguistic abilities.

It is, of course, possible that the primary function of language is something that is only needed occasionally. To be able to admonish some miscreant may be critical to the smooth running of a group, but it may not be necessary to do this more than once in a month of Sundays. That the machinery remains unused the rest of the time may be a cost worth bearing if the benefit is large enough. This would imply that all the wittering we do, all that social gossip, was just a matter of keeping the machinery of speech in trim, oiled and ready for the unpredictable moment when it suddenly becomes essential.

It seems like a plausible suggestion, but the enormous cost of having all this machinery in place seems rather excessive. Evolution is not usually so wasteful of resources, never mind time. Remember, the brain uses a fifth of all the energy of the body, about ten times more than you would expect on the basis of its size alone. Moreover, there are much simpler solutions to the problem of free-riders that are much more cost effective. Why don't we simply resort to the tactic that seems to work so well for all the other monkeys and apes – giving miscreants a jolly good thump? It would seem to be a great deal less costly in terms of brain size. In short, however valuable the policing function may be, it's unlikely to have been the precipitating cause of the evolution of large brains or language. Indeed, it seems logically wrong: the free-riders problem is a consequence of the fact that we live in large groups, and living in large groups seems not to be possible without big brains and language.

The suggestion that self-publicity may be an important consideration remains a more serious possibility. In fact, it is clear from our analyses of people's conversations that we do sometimes exploit the opportunities for advertising that language offers. Two particular findings from our studies point strongly in this direction.

One of the surprises of our study was that we found very few differences between males and females in the topics they talked about. Both sexes spent just as much time discussing personal relationships and experiences, and – contrary to popular myth – both spent time discussing other people's relationships and behaviour. Nor were the men in our samples more likely to discuss politics or high art (or for that matter low art) than the women were. There was, however, one striking difference: the proportion of time spent talking about work and academic matters or religion and ethics increased dramatically when males were in mixed sex groups. In each case, the proportion of total conversation time devoted to these topics increased from around 0–5 per cent in all-male groups to 15–20 per cent in mixed sex groups, with males showing a much more dramatic change in this respect than females.

Our interpretation of this result was that conversations often function as a kind of vocal lek. Leks are display areas where males gather to advertise their qualities as potential mates to the females. They occur widely among animals like antelope and birds, though usually only in species where the male does not (or cannot) contribute to the business of rearing the young. The peacock is a very typical lekking bird. The males defend small territories in an area that females frequent and display for all they are worth whenever any females come near them. The females wander from one male to another, assessing the qualities on offer. Eventually, having settled on the best of what may well be a bad bunch, each female mates with the male of her choice and then heads off elsewhere to get on with the business of laying eggs and rearing chicks at her leisure.

The suggestion that lekking may be an explanation for much of what passes as conversation in humans was given added weight by a second result from our conversation studies. We had found no differences between men and women in the amount of time devoted to talking about social topics: in both cases, around 65 per cent of speaking time was taken up with talking about social experiences of one kind or another. However, there was one respect in which they did differ, and that was *whose* social experiences they talked about most. At least among the younger group of subjects, the women tended to devote about two-thirds of their social topic time to other people's social experiences and activities (and about a third of the time to their own), whereas the men spent two-thirds talking about themselves (and only a third talking about other people).

This difference between the two sexes has important implications for what we might think is going on when people converse with each other. The most plausible interpretation is that the women are engaged in networking, while the men are engaged in advertising.

In terms of creating the right kind of environment for the successful rearing of offspring, networking is probably the single most important activity that women engage in. The support networks that these provide make it possible for the reciprocal exchange of information on the processes of childbirth and child rearing, help with foraging and horticulture, psychological support during times of emotional crisis and a dozen other support services large and small.

The male world, in contrast, is more directly competitive and much less co-operative. Directly or indirectly, much of its focus lies in mating or in the acquisition of the resources or status that will create future opportunities for mating. Advertising becomes a crucial factor in that process.

In the intellectual world of the university, demonstrating your intellectual skills by showing off your knowledge of Kant or the Romantic poets, or by being able to explain yesterday's lecture on the second law of thermodynamics, may be quite acceptable hallmarks of competence and status. They mark you out as a cut above the rest, the obvious choice in the mating stakes. In that kind of environment, intellectual prowess is as appropriate a criterion of future status or earning power as being the best card player in a bridge club or the best musician in a music club. Knowledge, as it is so often said, is power.

The Eyes Have It

It has been said that as much as two-thirds of the meaning in the sentences we utter is in fact conveyed in the nonverbal signals that accompany speech. Much is given away by the unspoken body language that we exhibit, sometimes deliberately, sometimes unintentionally, and we seem to be very sensitive to these cues.

I was particularly impressed by people's sensitivity to these aspects of the environment while carrying out a study of vigilance behaviour. The study was concerned with the rates at which people checked the environment even while engaged in conversations. To do this, we had to choose a subject and record every glance out at their environment. Look-ups are often very subtle and to be sure of not missing any, we had to stare at the subject's eyes for periods of up to five minutes, counting each look-up as it occurred. We soon discovered that people often noticed that they were being stared at, even from the far side of a large hall. People constantly monitor what is going on around them, sometimes by rapid glances around the room, but often simply out of the corner of an eye. As a result, we had to alter the way we collected our data in order to avoid upsetting subjects or disrupting their natural pattern of looking up, since being aware that someone is watching you causes you to look up more frequently.

In fact, it is clear that we both use and monitor cues of this kind constantly in everyday life. Eye contact, in particular, seems to be especially important as a sign of someone's honesty, as well as their interest in us. In her song 'The Butterflies Have Gone Away', the country singer Helen Darling is concerned because her lover's 'lonely eyes don't follow me no more', recognizing in this the first subtle sign of a dying love affair. One of the first signs of the impending collapse of the relationship between Prince Charles and Princess Diana picked up by the paparazzi was the fact that they neither touched nor looked each other in the eye during public appearances. We set great store by eye contact. As the saying goes: never trust a man that cannot look you in the eye.

The use of eye contact is particularly important in the context of initiating new relationships, especially for women. Controlling situations that could develop undesirable consequences is one of many vitally important tasks. Studies of behaviour in singles bars reveal that, in these contexts at least, women exert a surprising amount of control over the business of courtship and mating. With a few conspicuous exceptions (and in the absence of a lot of alcohol), males are surprisingly reluctant to pursue interactions with women unless they receive tacit encouragement with cues like eye contact. The two most important signals in this respect are strong steady eye-contact and the so-called 'coy' signal, in which eye contact is held for just a second, then followed by a rapid look away accompanied by a slight smile or blush (often with a second fleeting look back out of the corner of the eye a moment later).

Reading the signals is, in fact, a very ancient primate habit. This was demonstrated rather nicely in a study of hamadryas baboons by the Swiss zoologists Hans Kummer and Christian Bachmann. Unusually for baboons, male hamadryas hold small harems of females. Mating access to the females is jealously guarded by the harem-holder and males do not normally bother to challenge a harem male's hegemony over his females. Indeed, when near a harem, potential rivals will go out of their way to avoid giving even the slightest suggestion that they might be interested in the females: they will sit peering intently out into the distance, fiddling with grass blades and generally avoiding all eye-contact with either the male or his females.

Bachmann and Kummer found that, when the rival was very much more powerful than the harem male and knew from past experience that he could defeat the owner, he would sometimes attempt to take a female from the male. But he would only do so when the female was giving clear signals that she was not all that interested in her current male. The main cue seemed to be the alacrity with which the female followed the harem male when he moved and the frequency with which she glanced at him. Normally, hamadryas females will follow their males closely at heel. But her momentary hesitations, the dilitoriness that forces the male to stop and look back because she has not yet followed, are subtle cues that rival males read as showing that the female is less interested in her male than the conventional niceties of hamadryas society might normally require. The American primatologist Barbara Smuts noticed a very similar phenomenon in the olive baboons she studied in Kenya. Even though olive baboons are more promiscuous and informal in their relationships than the hamadryas, the males seemed to rely on the same subtle cues to determine whether a female was really interested in her current male consort.

But relieving a rival of his mate is not a simple matter. The mate in question needs to be persuaded that the change of male is worthwhile: when all else is equal, the devil you know is invariably better than the devil you don't. Prospective mates need to be impressed by a male's fitness to be their partner. Much of the business of mate selection, in humans as in other mammals, ends up with males advertising their wares and females choosing among those on offer. Finding a mate is, at root, a game of advertising.

The American anthropologist Kristin Hawkes has argued that hunting is just such a form of 'showing off' in traditional hunter-gatherer societies. She calculated the energy returns from hunting large game like antelope and concluded that it just isn't worth the energy and time expended on it. When the men do catch something, they promptly bring it back to camp and conspicuously share it with everyone else. In fact, they would do much better by setting out a dozen traps that they could visit every other day for five minutes. Yet the men insist on the importance of hunting and devote a great deal of time and effort to it, even though it makes no economic sense.

Hawkes argues that our mistake is to think of hunting as an economic activity related to parental investment – the archetypal prehistoric man going out hunting to feed the wife and kids. In fact, she suggests, it is really part of the mating game. Hunting large mammals is a difficult and risky business, and great skill is needed to pull off a kill successfully. A hunting male – and in most hunter-gatherer societies, men commonly hunt on their own or in groups of two or three – inevitably exposes himself to risk of ambush by predators like lions, as well as such dangers as snakes and elephants. Among the Eskimo in the days before snowmobiles made hunting less dangerous, males suffered extraordinarily high mortality rates out on the winter ice: a male's life expectancy could be less than half that of a woman's in some of the more extreme habitats of the Arctic north. In addition to these risks to his person, a hunting male has to exercise considerable skill in tracking and stalking prey animals in order to have a chance of a kill. As a demonstration of courage, stamina and skill, hunting provides infallible proof of how good your genes are.

Hunting has all the hallmarks of the challenges set to aspiring young knights in the chivalric tales of medieval Europe. In these stories, the young knight is given a superhuman task to test his mettle – rescue a damsel in distress, wake the sleeping beauty, kill the dragon that has been terrorizing the village, find the Holy Grail, fight the knight who has never been defeated, extract the sword from the stone. Such challenges are by no means confined to the folktales of Europe. Young Maasai warriors volunteer to throw away their spears and act as bait for a cornered lion. By walking in towards it holding his shield before him, the warrior forces the lion to jump him; his companions can then spear the lion in relative safety. By the time they have done this, however, the lion's hind claws – scrabbling for purchase beneath the young man's shield – have done their best to disembowel him. If he survives, he is feted as the hero of the village and becomes a much sought-after prize among the girls looking for husbands.

In a rather spectacular example of the same phenomenon, the young Captain Ewart Grogan walked the 4,500-mile length of Africa from the Cape of Good Hope to Cairo in 1899 to gain the hand of the woman he loved. Her family had dismissed him as a ne'er-do-well who would be unable to keep their daughter in the manner to which they thought she should be accustomed. Grogan banked on the fame (if not the fortune) that a dramatic adventure would bring him to persuade them to reconsider. They were duly impressed.

In our own societies, young males take risks racing cars at excessive speeds and play sports with a commitment and determination that few women would consider worthwhile. Yet, despite showing much less interest in performing these activities themselves, women continue to remain impressed by the performances of the males and compete with scarcely less vigour to associate with the champions of the game. The American basketball player Magic Johnson was no exception in being overwhelmed by the women who wanted to sleep with him. The fact that these activities are largely male mating displays may explain why, when women attempt them (walking Cape to Cairo or sailing single-handed around the world, to take two recent examples), it attracts a great deal less attention.

What all these 'heroic tasks' have in common is that they are difficult to cheat. They sort the men from the boys, those who are all mouth from those who genuinely can do. As a demonstration of fitness to be the father of your children and, perhaps more importantly, to provide for them in a capricious and uncertain world, it is difficult to imagine any more exacting test than hunting in the world of our ancestors.

The cognitive scientist Geoff Miller has thrown another light on the question by suggesting that the evolution of the human brain was driven mainly by the demands for sexual advertising. The ability to entertain a prospective mate, to charm them with poetry and song, to make them laugh – these, he suggests, are the things that the human brain has been designed to do. And it is not just a matter of catching your catch, for he or she can at any time become more impressed by someone else. Like the Red Queen whom Alice encountered in *Through the Looking Glass*, you have to keep running just to stay still. Just as the hunter has to keep hunting to prove that he is still the best catch in the community, so the modern male has to keep his partner smiling.

What makes this idea particularly appealing is a property of smiling and laughter that few people know about: they are both particularly good at stimulating the production of endogenous opiates. Both involve unusual muscle movements, while laughter, in particular, is surprisingly expensive in energetic terms. A bout of raucous laughter leaves us exhausted and gasping for breath. The staccato pumping of air through the windpipe requires a great deal of control and a lot of effort. Being morose and down-at-mouth makes you unhappier. So the best recipe for happiness in life is to smile as much as possible – thanks to the surge of opiates flooding through your veins, it makes you feel warm and contented. By the same token, making a prospective mate laugh lulls them into a sense of narcotic security.

Smiling and laughing have their own fascinating natural history. American psychologist Bob Provine recorded the frequency with which speakers and listeners laughed during conversations. He found that women are more likely to smile and laugh than men, that they are more likely to do so when they are listeners than when they are speaking, and that they are more likely to laugh in response to male speakers than to female speakers. By comparison, men are much less likely to laugh at something a woman says than they are at something a man says.

These are interesting findings for several reasons. One implication is that women comics are likely to be less successful than male comics because they will find it more difficult to raise a laugh from both male and female audiences. Women comics may have to act in a more exaggerated way than male comics do, breaking more strongly with conventional stereotypes for how men and women should behave.

These sex differences in smiling and laughing responses have been interpreted as reflections of the way society is dominated by males: women smile and laugh at men more because smiling and laughing express subordination. These behaviours are presumed to be the human equivalent of animal appeasement patterns like cowering with your tail between your legs.

Certain forms of 'false' smile or laugh may indeed be forms of appeasement in some circumstances. One study of doctors in a hospital, for example, showed that junior doctors were much more likely to smile at their seniors than vice versa, and were much more likely to laugh at their superiors' jokes. But there are many different kinds of smiles and laughs and not all of them are appeasement gestures. After all, we spend a great deal of time laughing and smiling at utterly helpless babies, as well as with our friends, without suffering from any sense of inferiority.

A much more plausible explanation is that women smile at men to encourage them to take an interest. They are engaged in a constant game of assessment, comparing the current partner with other males that hove into view. Most of the time they are happy to settle for what they already have, but it's important to keep testing the waters (after all, no one is perfect and you never know when your current mate might desert you). Testing a male's ability to make you laugh may be as good a way of covertly assessing his qualities as any.

The Mating Game

The pervasive importance of mate choice and sexual selection in our lives is illustrated by another well documented but surprising observation: men and women differ rather strikingly in the way they learn accents. As they mature, boys tend to adopt their local regional working class accents, whereas girls tend to pick up a more neutral middle-class standard form of English (known technically as Received Pronunciation, or RP for short). This curious fact has puzzled social scientists for years, since there is no obvious reason why the two sexes should differ in this way. The conventional explanation is that there is more social pressure on daughters to 'speak nicely': girls are more likely to be told they sound 'common' than boys are. It seems to be yet another example of the double standard where the boys get away with murder while the girls pay the price.

However, this is at best only half an explanation: it doesn't really tell us *why* these very different pressures should be exerted on the two sexes. So what is going on here? The answer is obvious once you appreciate that much of what we do (especially during the early adult years) is related to the business of mating and mate choice. The difference between the sexes reflects crucial differences in their respective reproductive strategies.

The main constraint on a female mammal's ability to reproduce is the resources she has available for rearing her offspring. Humans are no different, and women in almost all cultures show a striking preference for marrying males who are relatively wealthy (or high status, which comes to much the same thing as far as the opportunities of life are concerned). Jane Austen's novels of social life during the opening years of the nineteenth century illustrate this rather nicely, showing the young middle-class women of the day holding out for the ideal catch. The sons of the local vicar are rarely the objects of their attentions, but the dashing young army officers (conventionally the finishing school for the nobility or a passport to wealth and success for the middle classes) and the sons of landed gentry were much sought after. Unfortunately, of course, there are never enough of them to go round, and the girls cannot afford to wait for ever for fear of being left 'on the shelf'. Eventually some of them have to settle for second best – luckily for the local vicar's sons.

It is a surprise to most people to find that you still see much the same pattern in modern western societies. We have carried out three studies of mate choice preferences in contemporary society (one in the USA in collaboration with David Waynforth, and two in England). To do this, we analysed personal advertisements in newspapers and magazines, because these provide a neat encapsulation of what people would ideally like in a partner. As many as a quarter of the women mention cues of wealth and status – 'professional', 'home owner', 'college-educated', 'independent means' – as being desirable in a male partner, and 60–70 per cent of men mention these cues in writing about themselves, whereas women seldom mention them as descriptions of themselves and men seldom ask for them in women.

Given the fact that wealth and status are concentrated up the social ladder, it seems obvious that women will greatly improve their options in the marriage stakes by speaking in a way that will allow them to fit into the social world of the classes above them more easily. Hypergamy (or marrying up the social scale) is widespread in almost all human societies, and the behaviour patterns portrayed in Jane Austen's novels extended far beyond the confines of English county society. In rural Friesland in northwest Germany, detailed analyses of parish marriage registers from the last two centuries carried out by Eckart Voland (now at the University of Giessen) and his colleague Claudia Engel show that women tended to marry up the social scale whenever they could; marriages into the social class above were much more common than marriages into the social class below.

The wealthier petty farmers seem to have been highly sought after catches – and with good reason: the wealth they had to offer (small though it was in absolute terms) was enough to ensure significantly higher survival rates for their children. Moreover, women who married up the social hierarchy typically did so at a younger age than those who married males from their own social class. Even though most girls would eventually be forced to marry into their own social class, it was always worth hanging on that little bit longer just in case a better catch came along. But they could not afford to wait for ever, lest they miss out on the marriage stakes altogether. Bear in mind that we are not talking about Jane Austen's upper classes here, but about an essentially peasant society.

Marrying up the social (or economic) scale is still common today. This doesn't mean that every working-class girl marries an upper-class boy, but it does mean that marrying into the class above is more common among women than it is among men, and it is more common among women than marrying down the social scale. The daughter of an earl marrying her local dustman attracts far more attention than does the earl's son and heir marrying the dustman's daughter. Given this, it pays girls to develop an all-purpose accent that allows them to move more easily up the social scale when the opportunity arises – or at least it pays their parents to encourage them to do so.

The boys, on the other hand, face a rather different problem. Middle- and upper-class boys are in demand because they hold out the best hopes of being able to provide an adequate resource base on which to raise children; as a result, they often have to try less hard in the mating game. Lower-class boys, on the other hand, have less to offer in this respect; and because they are much more dependent on their community networks, it becomes important for them to ensure that they are seen to belong, that they have the right accents and dialects to mark them out as members of the group, where friends will provide them access to jobs or services that they would not otherwise be able to afford. Being poor with the wrong accent is the kiss of death, for it denies you access to the self-help network.

This emphasis on wealth and status has a simple utilitarian purpose. In society after society in the pre-industrial world, the single factor that most affects the mortality rates of children is the wealth of the husband. It doesn't seem to matter whether this comes in the form of land, cattle or money. A correlation between family resources and infant survival has been demonstrated among contemporary Kipsigis agro-pastoralists in Kenya by Monique Borgerhoff Mulder, in eighteenth- and nineteenth-century German peasant farmers by Eckart Voland, and in South American Ache hunter-gatherers by Kim Hill and Hilly Kaplan. The American psychologist David Buss has carried out a series of studies of mate choice preferences in some thirty-seven cultures drawn from all around the world: for the women, status and future earning potential were two of the most important criteria that they looked for in future husbands in virtually every culture. Having access to more wealth and resources enables you to feed your children better and, in modern economies, to create enough surplus to pay for more medical treatment and better education.

However, there is evidence to suggest that women's demands may be changing: in our samples of US and UK Lonely Hearts advertisements we found that around half the women were asking for family commitment in addition to or instead of wealth and status. This seems to be pointing to a shift in what women of reproductive age need for successful reproduction in modern economies. Where it was once resources, it is now increasingly the social input into the business of rearing children: help with child-care, a contribution to the socializing of the children. The contrast with the consistency of the preference for wealth and status in numerous studies of traditional societies is so striking that it cannot be an accident.

This change is of relatively recent origin (and a quarter of the women in our samples were still emphasizing wealth and status). It has come about, I think, because of two key changes that have taken place in modern industrial economies this century. One is that dramatically improved hygiene and health services have almost obliterated childhood mortality, so making it more or less certain that every child you give birth to will survive to adulthood. The other is that the general level of wealth is much greater, so that the differences between the wealthiest man in the community and the average one are no longer great enough to make the difference between a life of poverty and one that is good enough to provide for your children's needs. This second factor is, of course, compounded by women's own greater economic opportunities: they are no longer so dependent on their husbands to provide all the income for the household.

Men, it seems, have apparently not yet caught up with this shift in attitudes: the advertisements clearly show that they are still busily hammering away at the old virtues of wealth and status. Of course, real wealth still carries a lot of weight – witness the apparent ease with which millionaires of almost any age are able to attract beautiful young women. But the rest of us would probably do better to go for nappy-changing. Male behaviour will undoubtedly change, but it takes time; Eckart Voland and I have shown that, among rural German populations in the last two centuries, it took roughly a generation (thirty years) for rearing patterns to change in response to the precipitating economic factors that were driving change.

A further social implication of women becoming less dependent on wealth and resources for successful rearing is that there will be less need for hypergamy, less pressure to try to marry up the social scale. And if this happens, we should see an increasing tendency for girls to acquire their regional or class-based accents and less preference for RP.

Of course, these kinds of social change are predicated on the continued creation of sufficient wealth to allow all sectors of society to benefit. If economic slumps prevent us from achieving full employment and a wider distribution of wealth, mate choice patterns will return to the well-trod paths of yesteryear. Social change is driven by economic change.

Sitting at the hub of all this frenetic activity is the evolutionary mechanism known as sexual selection. The idea of sexual selection was first discussed by Charles Darwin over 120 years ago. He pointed out that some features in the natural world appear to have no survival benefits of any kind; indeed, they are very often counter-productive if measured purely in terms of their impact on an animal's survival. The classic case he had in mind was the peacock's tail: the male peafowl's long train makes its flight ungainly and awkward, and impedes its escape from pursuing predators. So why, asked Darwin, has the peacock's tail evolved? The answer, he suggested, is intense selection from peahens preferentially choosing males with the longest trains as mates. If the intensity of female choice is great enough, it will overpower the counter-selection being imposed by predators and other more mundane considerations like the energetic costs of flying while towing the equivalent of a barrage balloon.

Sexual selection has turned out to be a much more potent force in evolution than Darwin ever imagined. It may even be a more important force in the generation of new species than Darwin's original mechanism of environmentally driven natural selection. During the last thirty years a considerable amount of experimental and theoretical work has been done on this remarkable process, and we now know a great deal about it. The English biologist Marion Petrie, for example, has shown that peahens selectively prefer peacocks with the largest number of eye-spots on their tails. These males gain more matings, fertilise more eggs and have more surviving chicks than males with fewer eye-spots. These findings from real life were later confirmed experimentally by removing eyespots from some males' trains or by adding eyespots to the trains of other males. Those males who lost eyespots gained fewer matings than they had previously achieved, and those who gained eyespots had more matings. In another study, the Swedish biologist Malte Andersson demonstrated the same effect by reducing and lengthening the long trailing tail feathers of male widow birds in Kenya.

At least two mechanisms have been proposed to explain how this effect comes about. One is known as Zahavi's 'Handicap Principle' after the Israeli biologist Amotz Zahavi who first proposed it. Zahavi argued that what the males are, in effect, doing is saying: 'Just look at me! I'm *so* good that I can afford to weigh myself down with all this baggage and still outfly predators! Mate with me if you want sons and daughters as good as me!' This is Kristin Hawkes's showing off in another guise.

The other mechanism is known as Fisher's 'Sexy Sons Hypothesis'. The formidable English geneticist and statistician Ronald Fisher (one of the architects of the modern neo-Darwinian theory of evolution) suggested that female choice for quite arbitrary male characters might drive sexual selection sufficiently intensely to produce the kinds of useless characters like peacocks' tails. The point is a very simple one: if females happen to take a shine to a particular trait, such as eyespots, then their daughters are likely to inherit the same predisposition. Since it would benefit females to produce sons that have the characters that females prefer, it will pay them to mate preferentially with males with lots of eyespots (or whatever the character in question happens to be). This will lead to intense selection for males with many eyespots, and the rapid evolution of this trait in the male population.

In effect, Geoff Miller's poetic males hypothesis is a version of Fisher's Sexy Sons Hypothesis. Females who mate with males that carry these traits will produce sons that have these traits, who will then in their turn mate and produce lots of grandsons for their mothers. There is no intrinsic survival value to being a poet or a raconteur: it's simply a trait that females happen to have latched on to. Fisher's Sexy Sons Hypothesis is capable of producing very rapid evolution over relatively short periods of time. Which, of course, is just what we see in the evolution of the superbrains of modern humans: for nearly a million and a half years, brain volume remained roughly constant at around 700–800 cc, then in the space of just half a million years it all but doubled in size. And this, Geoff Miller argues, was all a consequence of intense selection for skills to keep your mate entertained.

There is another possibility, however, and it is suggested by the fact that smiling and laughing causes the brain to flood the body with endogenous opiates. Remember how, in Chapter 3, we saw that grooming is also very good at stimulating the production of endorphins? Think of the implications for bonding if the way grooming works is to make you feel very relaxed and mildly euphoric in the company of friends. If the intensity of a relationship is related to the amount of grooming effort put in (and hence to the quantities of opiates released), then our ancestors faced a serious problem when trying to push group sizes up beyond the levels observed in other primates. Leslie Aiello and I suggested that they initially did this by using vocalizations as a form of vocal grooming, so that they could keep 'grooming' with a friend even while busy feeding at a distance from them – just as the gelada in fact do today.

The problem here is that vocalizations are just, well , vocalizations. They don't have the same opiate-releasing properties as grooming. If opiates are a crucial part of the mechanism of bonding, it is likely that vocal exchanges will only allow you to increase the size of the group beyond the maximum for primates in general by a very limited amount. Sooner rather than later, you will hit an upper limit because the limited quantities of physical grooming cannot generate enough opiate production to maintain the reinforcement.

However, suppose that as language develops, signals associated with language themselves begin to stimulate opiate production. Smiling and, particularly, laughing do just this, and this may well explain why smiling and laughing are such important components of conversation. They may well have begun as signals of submission during the earliest phases. There are facial signals that are structurally very similar to both smiles and laughter in chimpanzees. But it seems that, at some point, they were captured and built into the business of bonding in social groups. We can now, quite literally, groom at a distance. Telling jokes allows us to stimulate opiate production in our grooming partners even when we don't have the time to sit there doing it physically. We can get on with the other important activities of ecological survival – travelling, hunting, gathering, preparing and eating food.

Looking back over the ideas we have explored in this chapter, it seems to me that what we have is not necessarily alternative hypotheses for the evolution of language and large brains in modern humans, but rather valuable new components that were added into the system as it developed. Once large brains and a language capacity had evolved as a way of bonding large groups, it opened up windows of opportunity in new directions. Deception and advertising were now possible, where they had not been before. They must surely have reinforced the processes of selection acting on large brains and improved language skills; they may even have driven these beyond the levels which would have been possible on the basis of social bonding alone. But without the conventional forces of social bonding to underpin them, they would not have been sufficiently powerful to drive the evolution of brain size upwards as rapidly as it actually occurred.

CHAPTER 10

The Scars of Evolution

Our journey has been a long and complicated one. We have covered five million or so years of evolutionary history; we have dipped into aspects of human biology as different as neurobiology and endocrinology on the one hand, and social psychology and anthropology on the other. Some of these will have struck a chord of familiarity. Others will have been new and surprising. So let me begin this final chapter by recapping the argument of the book.

The central argument revolves around four key points: (1) among primates, social group size appears to be limited by the size of the species' neocortex; (2) the size of human social networks appears to be limited for similar reasons to a value of around 150; (3) the time devoted to social grooming by primates is directly related to group size because it plays a crucial role in bonding groups; and finally, (4) it is suggested that language evolved among humans to replace social grooming because the grooming time required by our large groups made impossible demands on our time. Language, I argue, evolved to fill the gap because it allows us to use the time we have available for social interaction more efficiently.

Language fulfils this role in a number of different ways. It allows us to reach more individuals at the same time; it allows us to exchange information about our social world so that we can keep track of what's happening among the members of our social network (as well as keeping track of social cheats); it allows us to engage in self-advertising in a way that monkeys and apes cannot; and, last but not least, it seems to allow us to produce the reinforcing effects of grooming (opiate release) from a distance. For language to evolve, a number of key changes were required. Some of these were physiological (freeing off the energy required to maintain a large brain); others were cognitive (creating the brain modules needed to support ToM as well as the mechanical production of speech).

In this last chapter, I want to explore some of the implications of these findings for the way we live. I have borrowed my title from that of a book by Elaine Morgan in which she describes the many bits of the human body that are hangovers from our evolutionary past. Everything from our useless appendix to the weak backs we have through standing upright should remind us that evolution is not a process of inevitable perfection; in fact, evolution is a process of Heath Robinson make-do, a series of compromises by which we attempt to do the best we can with a set of incompatible goals. We are imperfect creatures, stuck with our evolutionary heritage, far from the perfection of design that eighteenth-century evolutionists interpreted as evidence for the handiwork of God.

The human mind is in no better shape than the human body. We are not quite Pleistocene minds trapped in space-age bodies, but there are some elements of our behaviour that reflect our evolutionary past and it probably is true that, in some cases at least, our cultural evolution has outstripped our ability to deal with the consequences.

This chapter, then, rather than presenting facts that we know to be true, comes more by way of speculation on what might be the case. The story I have told clearly has implications for much of what we do, but these implications have yet to be worked out in any detail. My aim is really to highlight some of the directions in which we might go.

Small is Best

Despite its extraordinary sophistication, human language is much more limited than we tend to give it credit for. Words fail us at crucial moments; we are unable to express the turmoil of inner thoughts that threaten to overwhelm us, so we resort to ancient modes of physical intimacy to express what we cannot or dare not say aloud. Of these limitations we are all painfully aware. But there are other respects in which the machinery of speech imposes quite serious constraints on how we talk to each other that are perhaps less familiar to us, even though we cope with some of them daily.

In Chapter 6, I pointed out that we are limited in the number of individuals whose attention we can hold in a conversation. Casual conversation groups are limited to around four people. This appears to be due to the fact that we cannot get more people into a circle small enough to hear what the speaker is saying. Two interesting features of our behaviour seem to arise from this.

One is that conversation groups never contain more than one speaker at a time. When they do, no one can keep track of the conversations, and the group either breaks up into two separate conversations or one speaker tries to dominate the other – either by speaking more loudly or by actually demanding silence from the others.

During our studies of conversational behaviour, it became apparent that there are striking sex differences in who does the talking in conversations that involve members of both the sexes. It has often been noted that, in mixed-sex groups, women tend to listen while men speak, and this has sometimes been interpreted as domineering behaviour on the part of men intent on forcing women into subservience. However, it is clear from our work that this explanation cannot be right (or at least not wholly right), for women are not more likely to listen on all occasions. In fact, the proportion of time that women spent speaking in male-female dyads (or pairs) is exactly 50 per cent when the dyad is the entire group, but it declines as the size of the group within which the conversing dyad is embedded increases. In groups of eight to twelve, a woman engaged in a 'private' conversation with a man is likely to be talking for only 25 per cent of the time.

There are two likely explanations for this. One is that, because women's voices are lighter than men's, they find it more difficult to be heard when the hubbub of conversation around them increases as the group they are in gets larger. When your efforts at conversation too often result in 'I'm sorry, I didn't catch that, could you say it again?', it's easier to sit back and listen. Since men's deeper voices carry further, it is inevitably the men that get left to do the talking.

There is, however, an alternative possibility. Since much of conversational behaviour among young adults has all the hallmarks of a mating lek (a display area on which males advertise their qualities so that females can choose among them), it stands to reason that when there are lots of men in the group, women should prefer to sit back and assess the bids on offer. You cannot assess the competition if you spend all the time talking – in fact, since speaking is a complex business, you probably won't have time to assess anything other than your own performance. In contrast, when a conversational dyad is on its own (rather than being embedded in a larger group), it is usually there for a very good reason: things have advanced from pure sales pitch to an attempt to build a relationship.

A second respect in which the mechanisms of speech impose constraints on our conversational behaviour concerns the way we handle groups that are larger than normal. In order to prevent Babel breaking out in committees or lecture halls, we have to impose very strict social rules on how people behave in these situations. In sermons or lectures, the bulk of the people present have to agree to suspend their right to speak in favour of one particular individual. This agreement is very fragile: if irate members of the audience really want to, they can easily prevent a speaker from continuing. In extreme cases, it may become necessary to evict troublesome members of the audience if business is to proceed at all.

These negotiated arrangements override the natural patterns of human behaviour in order to permit some broader collective advantage to emerge above the disruptive consequences of immediate self-interest. In effect, they are instances of what biologists call reciprocal altruism: this form of 'I'll scratch your back if you scratch mine' arrangement is another of the biological mechanisms by which altruistic behaviour can evolve in a Darwinian world. But, like all such forms of cooperation, it is open to invasion by cheats who may gain the benefits of cooperation without paying the price. By capitalizing on everyone else's willingness to adhere to the rules and remain silent, free-riders (for that is just

what they are) can ensure that they get heard by forcing themselves on the assembled company, shouting louder than anyone
else or intervening more vigorously.

Formal arrangements in which the audience suspends its right
to speak are clearly necessary to allow certain kinds of important
social functions to occur at all. Among the many such functions
that would otherwise become impossible in traditional societies
are religious instruction, rabble-rousing, proceedings in courts of
law and the more formal kinds of political decision-making. Even
things as simple as negotiating the arrangements for a marriage
would become impossible if everyone present at these events
insisted on speaking at the same time.

We encounter this problem in a particularly acute form in committees, where everyone present expects to be able to make a contribution. Committees require an effective chairman to exercise
control over the members, allowing each to speak in turn and preventing unnecessary interruptions. As anyone who has sat in committee meetings knows only too well, the chairman's control of
the meeting is fundamental. The moment the chairman's attention
is distracted, the committee very quickly breaks up into a set of
small conversations, often about anything except the business in
hand (and as often as not, comprising gossip about mutual
acquaintances).

The difficulty of controlling conversational interactions has
been largely responsible for an important informal rule in committee formation. It is a well-established principle that if you want
business done and real decisions made quickly, then you should
have a committee of no more than six members; if you want a
committee to do some brainstorming in order to generate some
new ideas, then you need more than six people in your committee.
The two functions seem to be wholly incompatible.

The bigger the committee, the longer it will take to come to any
conclusions. Too many people will want to have a say – there will
be too many *ifs*, *buts* and *howevers* – and numerically powerful
factions with opposing views may emerge. A small committee, on
the other hand, lacks a sufficiently large range of opinions to generate new ideas: it is more likely to come to a decision because

everyone has had a chance to have their say and there is nothing more to add. And if two radically opposed views do emerge, there won't be enough people to create sizable factions in support of both; consequently, it is easier to isolate someone who proposes an alternative view. Without the benefit of moral support, the odd man out is likely to accept the majority decision.

This might, indirectly, give us an explanation for the peculiarly paternalistic behaviour of the Victorians. Thanks to the success of their primary health programmes and the rapid developments in medical science, the Victorian middle and upper classes achieved previously unheard-of levels of infant survival. Family sizes increased from the 2–4 surviving offspring characteristic of most traditional peasant societies to the 4–8 typical of the better-off classes during the nineteenth century. With something like six children and two parents at the dinner table, plus the odd maiden aunt, the noise must have been deafening. It seems to me no surprise at all that the Victorians should have taken the view that children 'should be seen and not heard', and that they should speak only when spoken to. It was surely the only possible response in the face of what would otherwise have been bedlam. In contrast, with our 2.4 children and two adults, we can anticipate a more genteel discussion at the dinner table, and so treat children in a more liberal fashion. An obvious implication is that the authoritarian paternalism of the Victorians was an inevitable extension of domestic attitudes into the wider world of adult life.

One final example of the way our mental machinery seems to limit what we do concerns our attempts to create virtual conferencing systems. The technology for setting up conferencing calls in which several people are hooked up together through the telephone system has been available for some time. The telephone chat-line is one derivative of that technology. As a spin-off from that, much effort is now being put into building the technology for virtual conferencing systems, using video links to allow people in different parts of the world to work together on the same document or to discuss policy issues affecting a multinational company. This is obviously a much cheaper (and less exhausting) solution to the management problems of multinationals than flying people half-way round the globe for a two-hour meeting.

Unfortunately, it seems that the same kind of restrictions are likely to apply: it will be difficult to get more than about four people to interact successfully in these systems. The technology can handle an almost unlimited number of people, but the people cannot. Once there are more than four of them interacting, someone always gets left out and their contribution to the discussion becomes progressively marginalized.

It's as though our mental machinery for handling a conversation group has been set at the maximum number of people that we normally need to integrate into an interacting whole. Even though electronics can make it seem as though a dozen people are sitting next to each other, we just don't have the cognitive machinery necessary to keep more than about three of them in mind at the same time.

This has serious practical implications in areas like education. Most governments are predisposed to promote the development of large classes in order to minimize the costs of publicly funded education. But there is a cost. With large classes, the usual teaching strategy is a form of lecture, because the hubbub of voices created by any other form of teaching·is too disruptive. In a university context, much teaching is designed to try to provoke discussion, to get students to learn how to argue a case, to think through issues as they go along, weighing the evidence for and against alternative hypotheses or courses of action – but that is only possible with very small groups (six students and a teacher is usually considered the maximum viable group size for this kind of teaching). If the size of the group increases significantly beyond this, the discussion often becomes dominated by just a few of those present; the rest drift quietly into the metaphorical corners of their minds and gain little, or may set up competing conversations of their own.

What you can teach as well as how you teach is affected. In large classes, the exercise can become one of simply shovelling prepackaged information down open mouths. There is very little else you can do. The direct personal involvement in the to-and-fro of argumentation – the very thing that actively fires the mind – is lost, because the teacher's attention span is limited. The child's naturally questioning mind is forced to remain silent. And with that, the quality of education slips downwards from training minds that can think for themselves to training technicians who know the right response to give in a particular situation (but don't really know why it's the right one). Education comes to be a process of rote-learning rules of behaviour.

Coronation Street Blues

I wonder how many of us can honestly claim that we have not, at some point in our lives, become hooked on one or other of the TV serials. Bad as they often are, very few people escape them. Aside from our intrinsic fascination with other people's doings, there is an interesting issue here about why this particular form of entertainment should be so popular.

One of the more peculiar features of modern urban life is the extent to which we are locked into the tiny world of our own homes. Separated from relatives and with limited opportunities to create circles of friends, the modern city dweller is forced increasingly to draw on the ready-made imaginative family of the soap opera for a social life and a sense of community. It is conspicuous that the largest audience for these programmes is found among housebound women, trapped at home by young children. Those with active social lives, by contrast, rarely have an interest in these kinds of programmes.

No one has looked at this in any detail yet, but it won't surprise me if, when they do, they find that soap opera characters begin to fill the actual role of network members for those whose real social networks are kept well below the natural limit of 150 by their social or economic situation. Even TV newscasters and personalities can come to fill this role: they become part of our social networks, half-real friends whom we feel we know not just because we see them so often but also because they actually speak to *us* as individuals when they present the news. In fact, many successful newscasters deliberately try to speak in a way that makes it seem as though they are speaking to the individual listener across the dining room table.

In traditional peasant communities the world over, everyone lives in everyone else's pocket. They have to, of course, because houses are crammed together and walls are like paper. But more than that, people want to: the community is a genuine community, a co-operative whose members share the same problems of day-to-day survival. They are also bound by ties of kinship, at the very least through one sex and often through both.

Modern industrial conurbations often lack that sense of community because they have been created anew out of nothing: housing is built, and people trickle in from many different places to fill up the plots. They have no social ties, no common history to bind them. Their networks of friendship and kinship may be stretched far out beyond the confines of the housing estate, a problem that is exacerbated by the high rates of mobility that force people to move long distances in search of work.

One important consequence is that social networks become fragmented. In traditional societies, both peasant and hunter-gatherer, communities are tightly integrated units. Everyone shares the same wider network of acquaintances, everyone knows everyone else. Two individuals may not have the same circle of immediate friends and relatives (the dozen or so people with whom they interact most often), but their wider networks of 150 friends, relatives and acquaintances overlap almost completely. In post-industrial societies, this is virtually never the case. You and I share the same subset of acquaintances at work, but our spouses may not. You and your spouse may share another subset of acquaintances by virtue of belonging to the same church, but I do not. Rather than having a single large shared network, we have sets of sub-networks that only partially overlap. Each of us still has 150 people in our individual networks, but we may share only 15–20 members in common.

Our ties of common interest are weakened. By cooperating with you, I gain only from my immediate self-interest and by the benefits that you return to me in due course. In traditional communities, that benefit reverberates around the community in a series of overlapping waves as you pass on the benefit you received from me to your aunt, who in turn passes it on to her cousin, who passes it on to his friend – who eventually passes it back to me. My momentary generosity to you is repaid to me not once but repeatedly in the round-robin of social life in small communities. Despite the inevitable petty frustrations of life in small communities, the benefits of social obligation and reciprocation are magnified over and over again.

I am not suggesting that large towns are a bad thing or that people shouldn't move in search of work. Humans have been on the move since time immemorial. Large towns have been an economic and social magnet since at least the founding of the Middle Kingdom in Egypt and the heyday of the Mayan empire. Throughout the eighteenth century, London and other European capitals drew people in from far afield in search of work. They grew in size and power on the industrious backs of these immigrants. But what they had to offer was not always fame and fortune. Most of these cities grew despite mortality rates that exceeded their birth rates: they grew only because the number of immigrants continued to exceed the rate at which people died off.

Poor hygiene and low wages were a principal cause of mortality in the city slums. But these effects were unquestionably exacerbated by another factor that has been overlooked by demographers, and this was the lack of kinship and other support structures in migrant communities. The absence of kinship networks has a surprisingly bad effect on people's health. This was dramatically highlighted among both Captain Smith's Virginia colonists in 1626 and the famous Donner Party wagon train that set out to cross the American West in the 1846.

In both these cases, mortality was heaviest on those who had no relatives in the group. Despite often being fit young men at the outset, many of those who travelled alone with the Donner Party were unable to cope with the depredations of the journey. They died earlier and they died in significantly greater numbers. The same effect has been noted in a study of slum dwellers in northeastern England during the 1950s: those families with the smallest

kinship networks suffered the highest levels of both child mor-
bidity (sickness) and mortality, as well as being generally more
susceptible to conditions like depression. Networks of close social
ties seem to be fundamental to our survival. Similar results have
recently been reported from a study of a rural population in
Dominica.

This same loss of natural support networks seems to have been
responsible for the extraordinary rise in the number of religious
and pseudo-religious sects that have proved to be such magnets
for the young in the past half century. From Charles Manson to
David Koresh and the Rev. Chris Brain, from the Moonies to the
Hare Krishnas, it seems that it is the sense of belonging, of community, of family, that is of overriding importance in drawing in
young people. Indeed, some of the more aggressive of these
groups deliberately target lonely youngsters for this very reason.

In all these cases, beguiling language offers hope for a communal life that is more welcoming and more secure. Language preys
on the emotions, capitalizing on the fact that words can be used to
stir deep emotional feelings, to generate opiate highs when used in
the right way. History has many examples: religious fundamental-
ism sweeping whole nations, the rise of Fascism, witch-hunts,
pogroms and crusades all bear eloquent testimony to this process.
They are all consequences of the ease with which we are willing to
surrender our individuality to the collective will (or perhaps a single charismatic individual's will), spurred on by the hype of emotional speech. The psychological mechanism that evolved to facilitate the bonding of communities has lost its way because those
communities of common interest no longer exist. We are exposed
to the risk of exploitation by strangers. In the small community,
we have long-established bonds of trust, obligation and kinship to
guarantee that one person's fiercely argued view will not harm
others' interests. In the fragmented communities of the modern
world, we no longer have that guarantee. Yet the mechanisms that
engender trust in those who claim to be at one with us remain
firmly in place. Free-riders have never had it so good.

We can see this same effect spilling out in many different
aspects of our social lives. The extent to which Lonely Hearts

columns and dating agencies have boomed in the past two decades is indicative of the fact that people no longer have the kinds of social networks available that would normally provide them with access to prospective mates. The village match-maker has vanished with the village. As increasing numbers of people are thrown into a social vacuum through moving to a new city or town in pursuit of a job, more of us find ourselves in situations where we lack the social contacts needed to provide us with access to companions and partners. Where *do* you go to meet people without the risk of undesirable predators? Personal-ads columns and dating agencies are increasingly becoming a part of our normal social lives.

In somewhat similar vein, I am continually surprised at the extent to which friendships among adults in modern urban communities seem to stem not from the adults' own social contacts but from contacts established by their children through schools and clubs. It may not be too much of an exaggeration to suggest that improved nursery school provision may be more important for the parents than for the children.

This is not to say that there is anything intrinsically wrong with any of these phenomena, merely that they reflect the extent to which our psychological baggage may predispose us to certain kinds of social outlets. Nonetheless, the lack of social contact, the lack of a sense of community, may be the most pressing social problem of the new millennium.

Hard Sell around the Photocopier

It is often said that more business is conducted on the golf course than at the office desk. There is a very good reason why this should in fact be so. Business deals are personal interactions between individuals. The parties to a deal have to size each other up, assess the likelihood that the other party really means what they say, that they will stick to their word on a deal. You do not acquire that kind of information over the telephone or across a desk in a brief meeting. The golf itself is irrelevant: its purpose is simply to provide an opportunity for bonding.

The diamond dealers of New York and Amsterdam are in many ways an archetypal example of how a business community of this kind works. A man's word is his bond, because everyone in the community of dealers knows who he is, his past history, his honesty and reliability. The world of the gem dealer is a small, closed world of familiar faces and personal introductions. There is no need for contracts and documents. It all works by trust. But it can only work by trust because it is a small community. It will all collapse if too many people are allowed to join the club.

Contrast this with the amorphous super-networks of the international money markets. Large numbers of complete strangers are linked across the world through modern technology. How much of the current chaos of the money and insurance markets is a consequence of their size? Rogue dealers are able to get away with what they do because they are operating in a large anonymous market where obligation and trust do not exist, while at least some of their colleagues assume that they are still functioning in the small communities whose dealings are built on personal trust. In the modern dispersed electronic markets, dealers can never know all of those they come into contact with. Since trust between strangers is at best a fragile thing, behaviour will inevitably shift towards a new and less comfortable norm.

Proponents of the information superhighway have always hoped that access to networks of potentially unlimited size all around the world will open up wonderful opportunities for the mass communication of ideas: the global network at the forefront of information technology. Well, it's true that the flow of information will generally be greater: I can pick up things that someone I've never met (and probably never will meet) has deigned to make available on the Internet. But this won't necessarily open the doors to a worldwide network of associates and colleagues.

For one thing, the impersonality of the electronic highway seems to make people less discrete in their interactions with others than when they communicate face to face. They are more likely to be abusive when angry and more likely to make suggestive remarks in passing. What happens is somewhat akin to the 'road rage' with which we are becoming increasingly familiar.

Cocooned in their metal fortresses, people in cars escalate into anger much more quickly than they would had they been involved in an altercation as pedestrians on a sidewalk; cut off from direct face-to-face contact, where subtle cues are read rapidly and carefully, they lose the control that social interaction normally imposes in the interests of cooperation and bonding. Separated even further by the apparent anonymity of the computer link, there is even less to constrain us. The inevitable result will be 'Net rage'. Safe in the knowledge that our opponent cannot get at us, we feel confident about escalating fights we wouldn't dare risk in a car, never mind a face-to-face encounter.

Nor is it likely that electronic mail will significantly enlarge people's social networks. It may be faster than snail mail (as computer buffs refer to the conventional letter post), but it can have little effect on the human mind's ability to handle information about other people (as opposed to mere cyphers). The information superhighway's only real benefit in the end will be the speed with which ideas are disseminated. Whenever person-to-person interaction is a necessary feature of the process (as in the striking of deals), the old and trusted cognitive mindsets will come into play. Suspicion of the unknown and the fear of being duped by untrustworthy strangers will continue to dictate our decisions. As a result, negotiations in large amorphous populations are more likely to be conducted under the straitjacket of the rule book rather than by intuition. And where it is really important, we will resort to the trusted age-old machinery of direct personal contact. The old-boy and old-girl networks will never have seemed so important.

Sociologists have long recognized that businesses of less than 200 individuals can operate through the free flow of information among the members. But once their size exceeds this figure, some kind of hierarchical structure or line management system is necessary to prevent total chaos resulting from failures of communication. Imposing structures of this kind has its costs: information can only flow along certain channels because only certain individuals contact each other regularly; moreover, the lack of personalized contacts means that individuals lack that sense of personal commitment that makes the world of small groups go round.

Favours will only be done when there is a clear *quid pro quo*, an immediate return to the giver, rather than being a matter of communal obligation. Large organizations are just less flexible.

One solution to this problem would, of course, be to structure large organizations into smaller units of a size that can act as a cohesive group. By allowing these groups to build reciprocal alliances with each other, larger organizations can be built up. However, merely having groups of, say, 150 will never of itself be a panacea to the problems of organization. Something else is needed: the people involved must be able to build direct personal relationships. To allow free flow of information, they have to be able to interact in a casual way. Maintaining too formal a structure of relationships inevitably inhibits the way a system works.

The importance of this was drawn to my attention a couple of years ago by a TV producer. The production unit for which she worked produced all the educational output for a particular TV station. Whether by chance or by design, it so happened that there were almost exactly 150 people in the unit. The whole process worked very smoothly as an organization for many years until they were moved into new purpose-built accommodation. Then for no apparent reason, everything started to fall apart. The work seemed to be more difficult to do, not to say less satisfying.

It was some time before they worked out what the problem was. It turned out that, when the architects were designing the new building, they decided that the coffee room where everyone ate their sandwiches at lunchtime was an unnecessary luxury and so dispensed with it. The logic seemed to be that if people were encouraged to eat their sandwiches at their desks, then they were more likely to get on with their work and less likely to idle time away. And with that, they inadvertently destroyed the intimate social networks that empowered the whole organization. What had apparently been happening was that, as people gathered informally over their sandwiches in the coffee room, useful snippets of information were casually being exchanged. Someone had a problem they could not solve, and began to discuss it over lunch with a friend from another section. The friend knew just the person to ask. Or someone overhearing the conversation would have a suggestion, or would go away and happen to bump into some-one who knew the answer a day or so later; a quick phone call and the problem was resolved. Or a casual comment sparked an idea for a new programme.

It was these kinds of chance encounters over the coffee machine, idle chatter around the photocopier, that made the dif-ference between a successful organization and a less successful one. By encouraging casual contacts, the old system had created a network of relationships around each member of the staff and these acted like a parallel-processing supercomputer: several dif-ferent brains could be working on a problem at the same time independently of each other.

*

Language permeates human culture, underpinning our societies as much as our science and our art. Its roots go back into the distant past, and that ancient history is part of our mental baggage. We can do the most remarkable things with language. Yet underlying it all are minds that are not infinitely flexible, whose cognitive predispositions are designed to handle the kinds of small-scale societies that have characterized all but the last minutes of our evolutionary history.

That need not be a signal for despair. It is simply something we have to work with, something we have to adjust our social practices to take account of rather than fight against. Nor does it mean that human behaviour is incapable of change. That is a common but naive reading of the evolutionary message. Like all primates and many mammals, humans are characterized by behavioural flexibili-ty and the ability to adjust within the constraints of the machinery's design. The future of our species will be determined by our ability to recognize where those limitations lie and how we can circumvent them, if necessary by recreating the kinds of social environments in which we work best. If we can achieve that, the modern world may seem less alienating and become less destructive.

Bibliography

The following list includes the main journal articles in which the key results discussed in this book can be found; it is not intended as a complete reference list for all the studies quoted. For the general reader, a selection of books and articles that provide further reading on specific areas is included.

Chapter 1: Talking Heads
Milroy, R. (1987). *Language and Social Networks*. Blackwell, Oxford.
Lyons, J. (1970). *Chomsky*. Fontana, London.
Pinker, S. (1994). *The Language Instinct*. Allen Lane, London.
Tudgill, P. (1983). *Sociolinguistics: An Introduction to Language and Society*. Penguin, Harmondsworth.

Chapter 2: Into the Social Whirl
Bowler, P. J. (1986). *The Idea of Evolution*. University of California Press, Los Angeles.
Byrne, R. (1995). *The Thinking Ape*. Oxford University Press, Oxford.
Byrne, R., and Whiten, A. (1987). 'The thinking primate's guide to deception.' *New Scientist* 116 (No. 1589): 54–7.
Byrne, R., and Whiten, A. (eds.) (1988). *Machiavellian Intelligence*. Oxford University Press, Oxford.
Cheney, D. L., and Seyfarth, R. M. (1990). *How Monkeys See the World*. Chicago University Press, Chicago.
Cheney, D. L., Seyfarth, R. M., and Silk, J. B. (1995). 'The role of grunts in reconciling opponents and facilitating interactions among adult female baboons.' *Animal Behaviour* 50: 249–57.
Dawkins, R. (1976). *The Selfish Gene*. Oxford University Press, Oxford.
Dawkins, R. (1993). 'Gaps in the mind.' In: P. Cavalieri and P. Singer (eds.), *The Great Ape Project*, pp. 80–87. Fourth Estate, London.
Dennett, D. (1995). *Darwin's Dangerous Idea*. Allen Lane, Harmondsworth.

Dunbar, R. I. M. (1984). *Reproductive Decisions: An Economic Analysis of Gelada Baboon Social Strategies.* Princeton University Press, Princeton.

Dunbar, R. I. M. (1988). *Primate Social Systems.* Chapman & Hall, London.

Fleagle, J. G. (1988). *Primate Adaptation and Evolution.* Academic Press, New York.

Goodall, J. (1986). *The Chimpanzees of Gombe: Patterns of Behaviour.* Harvard University Press, Cambridge (Mass.).

Gribbin, J., and Gribbin, M. (1993). *Being Human: Putting People in an Evolutionary Perspective.* Dent, London.

Harcourt, A., and de Waal, F. (eds.) (1993). *Coalitions and Alliances in Humans and Other Animals.* Oxford University Press, Oxford.

Hinde, R. A. (ed.) (1983). *Primate Social Relationships.* Blackwells Scientific Publications, Oxford.

Jones, S., Martin, R. D., and Pilbeam D. (eds.) (1992). *The Cambridge Encyclopedia of Human Evolution.* Cambridge University Press, Cambridge.

Martin, R. D. (1990). *Primate Origins and Evolution.* Chapman & Hall, London.

Pearson, R. (1978). *Climate and Evolution.* Academic Press, London.

Richard, Alison F. (1985). *Primates in Nature.* W. H. Freeman, San Francisco.

Smuts, B. B., Cheney, D. L., Seyfarth, R. M., Wrangham, R. W., and Struhsaker, T. T. (eds.) (1987). *Primate Societies.* Chicago University Press, Chicago.

Smuts, B. B. (1985). *Sex and Friendship in Baboons.* Aldine, New York.

de Waal, F. (1982). *Chimpanzee Politics.* Allen & Unwin, London.

de Waal, F., and van Roosmalen, J. (1979). 'Reconciliation and consolation among chimpanzees.' *Behavioural Ecology and Sociobiology* 5: 55–66.

Chapter 3: The Importance of Being Earnest

Abott, D. H. (1984). 'Behavioural and physiological suppression of fertility in subordinate marmoset monkeys.' *American Journal of Primatology* 6: 169–86.

Abbott, D. H., Keverne, E. B., Moore, G. F., and Yodyinguad. U. (1986). 'Social suppression of reproduction in subordinate talapoin monkeys, *Miopithecus talapoin*.' In: J. Else and P. C. Lee (eds.), *Primate Ontogeny*, pp. 329–41. Cambridge University Press, Cambridge.

Barton, R. (1985). 'Grooming site preferences in primates and their functional implications.' *International Journal of Primatology* 6: 519–31.

Bowman, L. A., Dilley, S. R., and Keverne, E. B. (1978). 'Suppression of oestrogen-induced LH surges by social subordination in talapoin monkeys.' *Nature,* London, 275: 56–8.

Cheney, D. L., and Seyfarth, S. M. (1990). *How Monkeys See the World.* Chicago University Press, Chicago.

Dunbar, R. I. M. (1985). 'Stress is a good contraceptive.' *New Scientist* 105 (17 January): 16–18.

Dunbar, R. I. M. (1988). *Primate Social Systems.* Chapman & Hall, London.

Dunbar, R. I. M. (1991). 'Functional significance of social grooming in primates.' *Folia primatologica* 57: 121–31.

Enquist, M., and Leimar, O. (1993). 'The evolution of cooperation in mobile organisms.' *Animal Behaviour* 45: 747–57.

Goosen, C. (1981). 'On the function of allogrooming in Old World monkeys.' In: A. B. Chiarelli and R. S. Corruccini (eds.), *Primate Behaviour and Sociobiology,* pp. 110–20. Springer, Berlin.

Hockett, C. F. (1960). 'Logical considerations in the study of animal communication.' In: W. E. Lanyon and W. N. Tavolga (eds.), *Animal Sounds and Communication,* pp. 392–430. American Institute of Biological Sciences, Washington.

Howlett, T., Tomlin, S., Ngahfoong, L., Rees, L., Bullen, B., Skrinar, G., and MacArthur, J. (1984). 'Release of ß endorphin and met-enkephalin during exercise in normal women: response to training.' *British Medical Journal* 288: 1950–2.

Keverne, E. B., Martensz, N., and Tuite, B. (1989). 'Beta-endorphin concentrations in cerebrospinal fluid on monkeys are influenced by grooming relationships.' *Psychoneuroendcrin-ology* 14: 155–61.

Mason, H. (1984). 'Everything you wanted to know about sperm banks.' *Observer* (20 August), p. 35.

Silk, J. (1982). 'Altruism among female *Macaca radiata*: explanations and analysis of patterns of grooming and coalition formation.' *Behaviour* 79: 162–88.

Sparks, J. (1967). 'Allogrooming in primates: a review.' In: D. Morris (ed.), *Primate Ethology,* pp. 148–75. Weidenfeld & Nicholson, London.

Smuts, B. B., Cheney, D. L., Seyfarth, R. M., Wrangham, R. W., and Struhsaker, T. T. (eds.) (1987). *Primate Societies.* Chicago University Press, Chicago.

Wasser, S. K., and Barash, D. P. (1983). 'Reproductive suppression among female mammals: implications for biomedicine and sexual selection theory.' *Quarterly Reviews of Biology* 58: 513–38.

Chapter 4: Of Brains and Groups and Evolution

Buys, C. J., and Larsen, K. L. (1979). 'Human sympathy groups.' *Psychology Reports* 45: 547–53.

Bibliography

Byrne, R. (1995). *The Thinking Ape*. Oxford University Press, Oxford.

Byrne, R., and Whiten, A. (eds.) (1988). *Machiavellian Intelligence*. Oxford University Press, Oxford.

Coleman, J. S. (1964). *An Introduction to Mathematical Sociology*. Collier-Macmillan, London.

Dunbar, R. I. M. (1992). 'Neocortex size as a constraint on group size in primates.' *Journal of Human Evolution* 20: 469–93.

Dunbar, R. I. M. (1993). 'Coevolution of neocortical size, group size and language in humans.' *Behavioural and Brain Sciences* 16: 681–735.

Dunbar, R. I. M. (1992). 'Why gossip is good for you.' *New Scientist* 136: 28–31.

Dunbar, R. I. M., and Spoors, M. (1995). 'Social networks, support cliques amd kinship.' *Human Nature* 6: 273–90.

Friedman, J., and Rowlands, M. J. (eds.) (1977). *The Evolution of Social Systems*. Duckworth, London.

Hayes, K., and Hayes, C. (1952). 'Imitation in a home-raised chimpanzee.' *Journal of Comparative Psychology* 45: 450–9.

Jerison, H. J. (1973). *Evolution of the .Brain and Intelligence*. Academic Press, New York.

Johnson, G. A. (1982). 'Organizational structure and scalar stress.' In: Renfrew, C., Rowlands, M., and Abbott-Seagram, B. (eds.), *Theory and Explanation in Archaeology*. Academic Press, London.

Kellogg, W. N., and Kellogg, L. A. (1933). *The Ape and the Child: A Study of Environmental Influence upon Early behaviour*. Whittlesey House, New York.

Killworth, P. D., Bernard, H. R., and McCarty, C. (1984). 'Measuring patterns of acquaintanceship.' *Current Anthropology* 25: 391–7.

Kudo, H., Bloom, S., and Dunbar, R. I. M. (submitted). 'Neocortex size as a constraint on social network size in primates.' *Behaviour*.

Lewin, R. (1992). 'The great brain race.' *New Scientist* (5 December), pp. 2–8.

Naroll, R. (1956). 'A preliminary index of social development.' *American Anthropologist* 58: 687–715.

Premack, D., and Premack, A. J. (1983). *The Mind of an Ape*. Norton, New York.

Savage-Rumbaugh, S. (1980). *Ape Language: From Conditioned Response to Symbol*. Oxford University Press, New York.

Chapter 5: The Ghost in the Machine

Astington, J. W. (1994). *The Child's Discovery of the Mind*. Fontana, London.

Barkow, J. H., Cosmides, L., and Tooby, J. (eds.) (1993). *The Adapted Mind*. Oxford University Press, Oxford.

Baron-Cohen, S. (1991). 'The theory of mind impairment in autism.' In:

Grooming, Gossip and the Evolution of Language

Whiten, A. (ed.), *Natural Theories of Mind*, pp. 233–52. Oxford University Press, Oxford.

Byrne, R. (1995). *The Thinking Ape*. Oxford University Press, Oxford.

Cheney, D. L., and Seyfarth, S. M. (1990). *How Monkeys See the World*. Chicago University Press, Chicago.

Dennett, D. (1983). 'Intentional systems in cognitive ethology: the "Panglossian paradigm" defended.' *Behavioural and Brain Sciences* 6: 343–90.

Donald, M. (1991). *Origins of the Modern Mind*. Harvard University Press, Cambridge (Mass).

Dunbar, R. I. M. (1995). *The Trouble with Science*. Faber & Faber, London.

Gallup, G. G. (1970). 'Chimpanzees: self-recognition.' *Science* 167: 417–21.

Gallup, G. G. (1985). 'Do minds exist in species other than our own?' *Neuroscience and Biobehavioural Reviews* 9: 631–41.

Happé, F. (1994). *Autism: An Introduction to Psychological Theory*. University College London Press, London.

Kinderman, P., Dunbar, R., and Bentall, R. (submitted). 'Theory-of-mind deficits, causal attributions and paranoia: an analogue study.' *British Journal of Psychology*.

Leslie, A. M. (1987). 'Pretence and representation in infancy: the origins of theory of mind.' *Psychological Review* 94: 84–106.

Parker, S., Mitchell, R. W., and Boccia, M. L. (eds.) (1994). *Self-awareness in Animals and Humans*. Cambridge University Press, Cambridge.

Povinelli, D. J. (1989). 'Failure to find self-recognition in Asian elephants (*Elephas maximus*) in contrast to their use of mirror cues to discover hidden food.' *Journal of Comparative Psychology* 103: 122–31.

Povinelli, D. J., Nelson, K. E., and Boysen, S. T. (1990). 'Inferences about guessing and knowing by chimpanzees (*Pan troglodytes*).' *Journal of Comparative Psychology* 104: 203–10.

Povinelli, D. J., Nelson, K. E. and Boysen, S. T. (1992). 'Comprehension of social role reversal by chimpanzees: evidence of empathy?' *Animal Behaviour* 43: 633–40.

Premack, D., and Woodruff, G. (1978). 'Does the chimpanzee have a theory of mind?' *Behavioural and Brain Sciences* 4: 515–26.

RACTER (1985). *The Policeman's Beard is Half Constructed*. Warner Books, New York.

Weiskrantz, L. (ed.) (1985). *Animal Intelligence*. Oxford University Press, Oxford.

Wolpert, L. (1994). *The Unnatural Nature of Science*. Faber & Faber, London.

de Waal, F. (1982). *Chimpanzee Politics*. Allen & Unwin, London.

Whiten, A. (ed.) (1991). *Natural Theories of Mind*. Blackwell, Oxford.

Whiten, A. , and Byrne, R. (1988). 'Tactical deception in primates.' *Behavioural and Brain Sciences* 11: 233–44.

Chapter 6: Up Through the Mists of Time

Aiello, L., and Dunbar, R. I. M. (1993). 'Neocortex size, group size and the evolution of language.' *Current Anthropology* 34: 184–93.

Aiello, L., and Wheeler, P. (1995). 'The expensive tissue hypothesis.' *Current Anthropology* 36: 199–211.

Alexander, R. D., Hoogland, J. L., Howard, R. D., Noonan, K. M., and Sherman, P. W. (1979). 'Sexual dimorphisms and breeding systems in pinnipeds., ungulates, primates and humans.' In: N. Chagnon and W. Irons (eds.), *Evolutionary Biology and Human Social Behaviour*, pp. 402–35. Duxbury, North Scituate (Mass.).

Arensburg, B., Tillier, A. M., Vandermeersch, B., Duday, H., Schepartz, L. A., and Rak, Y. (1989). 'A Middle Palaeolithic human hyoid bone.' *Nature*, London, 338: 758–9.

Bischoping, K. (1993). 'Gender differences in conversation topics, 1922–1990.' *Sex Roles* 28: 1–17.

Clarke, R. J., and Tobias, P. V. (1995). 'Sterkfontein member 2 foot bones of the oldest South African hominid.' *Science* 269: 521–4.

Cohen, J. E. (1971). *Casual Groups of Monkeys and Men*. Harvard University Press, Cambridge (Mass.).

Dunbar, R. I. M. (1988). *Primate Social Systems*. Chapman & Hall, London.

Dunbar, R. I. M. (1993). 'Coevolution of neocortical size, group size and language in humans.' *Behavioural and Brain Sciences* 16: 681–735.

Dunbar, R. I. M., Duncan, N. D. C., and Nettle, D. (1995). 'Size and structure of freely forming conversational groups.' *Human Nature* 6: 67–78.

Dunbar, R. I. M., and Spoors, M. (1995). 'Social networks, support cliques and kinship.' *Human Nature* 6: 273–90.

Janis, C. (1976). 'The evolutionary strategy of the Equidae and the origins of rumen and caecal digestion.' *Evolution* 30: 757–74.

Jones, S., Martin, R. D., and Pilbeam D. (eds.) (1992). *The Cambridge Encyclopedia of Human Evolution*. Cambridge University Press, Cambridge.

Legget, R. F., and Northwood, T. D. (1960). 'Noise surveys at cocktail parties.' *Journal of the Acoustical Society of America* 32: 16–18.

Liebermann, D. (1989). 'The origins of some aspects of language and cognition.' In P. Mellars and C. Stringer (eds.), *The Human Revolution*, pp. 391–414. Edinburgh University Press, Edinburgh.

Mellars, P., and Stringer, C. (eds.) (1989). *The Human Revolution*. Edinburgh University Press, Edinburgh.

Nettle, D. (1994). 'A behavioural correlate of phonological structure.' *Language and Speech* 37: 425–9.

Pinker, S. (1994). *The Language Instinct*. Allen Lane, London.

Sigg, H., and Stolba, A. (1981). 'Home range and daily march in a hamadryas baboon troop.' *Folia Primatologica* 36: 40–75.

van Soest, P. J. (1982). *The Nutritional Ecology of the Ruminant*. Cornell University Press, Ithaca.

Stringer, C., and Gamble, C. (1993). *In Search of the Neanderthals*. Thames and Hudson, London.

Wheeler, P. E. (1988). 'Stand tall to stay cool.' *New Scientist* (December) pp. 62–5.

Wheeler, P. E. (1991). 'The influence of bipedalism on the energy and water budgets of early hominids.' *Journal of Human Evolution* 21: 107–36.

Chapter 7: First Words

Asch, S. E. (1956). 'Studies of independence and conformity. A minority of one against a unanimous majority.' *Psychological Monographs* 70, No. 9.

Bever, G., and Chiarello, R. J. (1974). 'Cerebral dominance in musicians and non-musicians.' *Science* 185: 137–9.

Bott, E. (1971). *Family and Social Network*. Tavistock Publications, London.

Bradshaw, J., and Rogers, L. (1993). *The Evolution of Lateral Asymmetries, Language, Tool Use and Intellect*. Academic Press, New York.

Calvin, W. H. (1983). *The Throwing Madonna: From Nervous Cells to Hominid Brains*. McGraw-Hill, New York.

Casperd, J., and Dunbar, R. I. M. (1996). 'Asymmetries in the visual processing of emotional cues during agonistic interactions by gelada baboons.' *Behavioural Processes* (in press).

Cheney, D. L., and Seyfarth, R. M. (1982). 'How vervet monkeys perceive their grunts: field playback experiments.' *Animal Behaviour* 30: 739–51.

Corballis, M. C. (1991). *The Lopsided Ape*. Oxford University Press, Oxford.

Davies, N. B., and Halliday, T. R. (1977).' Optimal mate selection in the toad *Bufo bufo*.' *Nature*, London, 269: 56–8.

Denman, J., and Manning, J. T. (submitted). 'Lateral cradling preferences and left-eye-mediated perceptions of emotions.' *Ethology and Sociobiology*.

Dunbar, R. I. M. (1988). *Primate Social Systems*. Chapman & Hall, London.

Bibliography

Dunbar, R. I. M., and Spoors, M. (1995). 'Social networks, support cliques amd kinship.' *Human Nature* 6: 273–90.

Foley, R. A. (1989). 'The evolution of hominid social behaviour.' In: V. Standen and R. Foley (eds.), *Comparative Socioecology*, pp. 473–94. Blackwell Scientific Publications, Oxford.

Foley, R. A., and Lee, P. C. (1989). 'Finite social space, evolutionary pathways and reconstructing hominid behaviour.' *Science* 243: 901–6.

Hauser, M. (1993). 'Right hemisphere dominance for the production of facial expression in monkeys.' *Science* 261: 475–7.

Hauser, M., and Fowler, C. (1991). 'Declination in fundamental frequency is not unique to human speech.' *Journal of the Acoustical Society of America* 91: 363–9.

Jaynes, J. (1990). *The Origin of Conciousness in the Breakdown of the Bicameral Mind.* Houghton Mifflin, New York.

Kinzey, W. (ed.) (1987). *The Evolution of Human Behaviour: Primate Models.* State University of New York Press, Albany.

Knight, C. (1990). *Blood Relations: Menstruation and the Origins of Culture.* Yale University Press, New Haven (Conn.).

Manning, J. T., Chamberlain, A. T., and Heaton, R. (1994). 'Left-sided cradling: similarities and differences between apes and humans.' *Journal of Human Evolution* 26: 77–83.

Richman, B. (1976). 'Some vocal distinctive features used by gelada monkeys.' *Journal of the Acoustical Society of America* 60: 718–24.

Rodseth, L., Wrangham, R. W. , Harrigan, A., and Smuts, B. B. (1991). 'The human community as a primate society.' *Current Anthropology* 32: 221–55.

Schumacher, A. (1982). 'On the significance of stature in human society.' *Journal of Human Evolution* 11: 697–701.

de Waal, F. (1982). *Chimpanzee Politics.* Allen & Unwin, London.

de Waal, F. (1984). 'Sex differences in the formation of coalitions among chimpanzees.' *Ethology and Sociobiology* 5: 239–55.

Wallach, M. A., Kogan, N., and Bem, D. J. (1962). 'Group influence on individual risk-taking.' *Journal of Abnormal and Social Psychology* 65: 75–86.

Wallach, M. A., Kogan, N., and Bem, D. J. (1964). 'Diffusion of responsibility and level of risk-taking in groups.' *Journal of Abnormal and Social Psychology* 68: 263–74.

Chapter 8: Babel's Legacy

Cavalli-Sforza, L. L., Piazza, A., Menozzi, P., and Mountain, J. L. (1988). 'Reconstruction of human evolution: bringing together genetic, archaeological and linguistic data.' *Proceedings of the National Academy of Sciences, USA* 85: 6002–6.

Daly, M., and Wilson, M. (1988). *Homicide*. Aldine, New York.

Dawkins, R. (1976). *The Selfish Gene*. Oxford University Press, Oxford.

Dunbar, R. I. M., Clark, A., and Hurst, N. L. (1995). 'Conflict and cooperation among the Vikings: contingent behavioural decisions.' *Ethology and Sociobiology* 16: 233–46.

Green, S. (1975). 'Dialects in Japanese monkeys: vocal learning and cultural transmission of locale-specific vocal behaviour?' *Zeitschrift fur Tierpsychologie* 38: 304–14.

Hughes, A. L. (1988). *Evolution and Human Kinship*. Oxford University Press, Oxford.

Johnson, G. R., Ratwick, S. H., and Swyer, T. J. (1987). 'The evocative significance of kin terms in patriotic speech.' In: Reynolds, V., Falger, V., and Vine, I. (eds.), *The Sociobiology of Ethnocentrism*, pp. 157–74. Chapman & Hall, London.

Knight, C. (1990). *Blood Relations: Menstruation and the Origins of Culture*. Yale University Press, New Haven.

Mitani, J., Hasegawa, T., Gros-Louis, J., Marler, P., and Byrne, R. (1992). 'Dialects in wild chimpanzees?' *American Journal of Primatology* 23: 233–43.

Nettle, D., and Dunbar, R. I. M. (submitted). 'Social markers and the evolution of reciprocal exchange.' *Current Anthropology*.

Panter-Brick, C. (1989). 'Motherhood and subsistence work: the Tamang of rural Nepal.' *Human Ecology* 17: 205–28.

Pinker, S. (1994). *The Language Instinct*. Allen Lane, London.

Renfrew, C. (1994). 'World linguistic diversity.' *Scientific American* 270: 104–11.

Ridley, M. (1994). *The Red Queen*. Viking, London.

Shaw, R. P., and Wong, Y. (1989). *Genetic Seeds of Warfare: Evolution, Natiuonalism and Patriotism*. Unwin Hyman, Boston.

Stoneking, M., and Cann, R. (1989).' African origin of human mitochondrial DNA.' In: Mellars, P., and Stringer, C. (eds.). *The Human Revolution*, pp. 17–30. Edinburgh University Press, Edinburgh.

Chapter 9: The Little Rituals of Life

Bachmann, C., and Kummer, H. (1980). 'Male assessment of female choice in hamadryas baboons.' *Behavioural Ecolology and Sociobiology* 6: 315–21.

Betzig, L., Borgerhoff Mulder, M., and Turke, P. (1988). *Human Reproductive Behaviour*. Cambridge University Press, Cambridge.

Bischoping, K. (1993). 'Gender differences in conversation topics, 1922–1990.' *Sex Roles* 28: 1–18.

Buss, D. (1994). *The Evolution of Desire*. Basic Books, New York.

Coates, J. (1993). *Women, Men and Language*. Longman, New York.

Coser, R. L. (1960). 'Laughter among colleagues.' *Psychiatry* 23: 81–95.

Cosmides, L., and Tooby, J. (1993). 'Cognitive adaptations for social exchange.' In: Barkow, J. H., Cosmides, L., and Tooby, J. (eds.), *The Adapted Mind*, pp. 162–228. Oxford University Press, Oxford.

Daly, M., and Wilson, M. (1988). 'Evolutionary psychology and family homicide.' *Science* 242: 519–24.

Dunbar, R. I. M. (1993). 'The co-evolution of neocortical size, group size and the evolution of language in humans.' *Behavioural and Brain Sciences* 16: 681–735.

Dunbar, R. I. M., Duncan, N. D. C., and Marriott, A. (submitted). 'Human conversational behaviour: a functional approach.' *Ethology and Sociobiology*.

Eakins, B. W., and Eakins, R. G. (1978). *Sex Differences in Human Communication*. Houghton Mifflin, Boston.

Emler, N. (1990). 'A social psychology of reputations.' *European Journal of Social Psychology* 1: 171–93.

Emler, N. (1992). 'The truth about gossip.' *Social Psychology Newsletter* 27: 23–37.

Goodman, R. F., and Ben-Ze'ev, A. (eds.) (1994). *Good Gossip*. University of Kansas Press.

Grammer, K. (1989). 'Human courtship behaviour: biological basis and cognitive processing.' In: A. Rasa, C. Vogel and E. Voland (eds.), *The Sociobiology of Sexual and Reproductive Behaviour*, pp. 147–69. Chapman & Hall, London.

Hawkes, K. (1991). 'Showing off: tests of another hypothesis about men's foraging goals.' *Ethology and Sociobiology* 11: 29–54.

Huxley, E. (1987). *Out in the Midday Sun*. Penguin, Harmondsworth.

Kipers, P. S. (1987). 'Gender and topic.' *Language and Society* 16: 543–57.

Miller, G. (1996). 'Sexual selection in human evolution: review and prospects.' In: C. Crawford and D. Krebs (eds.), *Evolution and Human Behaviour: Ideas, Issues and Applications*. Lawrence Erlbaum, New York.

Moore, M. M. (1985). 'Non-verbal courtship patterns in women: context and consequences.' *Ethology and Sociobiology* 6: 237–47.

Petrie, M. (1994). 'Improved growth and survival of offspring of peacocks with more elaborate trains.' *Nature*, London 371: 598–9.

Petrie, M., and Halliday, T. (1994). 'Experimental and natural changes in the peacock's (*Pave cristatus*) train can affect mating success.' *Behavioural Ecology and Sociobiology* 35: 213–17.

Provine, R. R. (1993). 'Laughter punctuates speech: linguistic, social and gender contexts of laughter.' *Ethology* 95: 291–8.

Ridley, M. (1994) *The Red Queen*. Viking, London.

Smuts, B. B. (1985). *Sex and Friendship in Baboons*. Aldine, New York

Voland, E., and Engel, C. (1990). 'Female choice in humans: a condition-al mate choice strategy of the Krummhorn women (Germany 1720–1874).' *Ethology* 84: 144–54.

Waynforth, D., and Dunbar, R. I. M. (1995). 'Conditional mate choice strategies in humans: evidence from lonely-hearts advertisements.' *Behaviour* 132: 755–79.

Zahavi, A. (1975). 'Mate selection – a selection for a handicap.' *Journal of Theoretical Biology* 53: 205–14.

Chapter 10: The Scars of Evolution

Bott, E. (1971). *Family and Social Network*. Tavistock Publications, London.

Cohen, J. (1971). *Casual Groups of Monkeys and Men*. Harvard University Press, Cambridge (Mass.).

Coleman, J. S. (1964). *Introduction to Mathematical Sociology*. Collier-Macmillan, London.

Dunbar, R. I. M., Duncan, N. D. C., and Nettle, D. (1995). 'Size and structure of freely forming conversational groups.' *Human Nature* 6: 67–78.

Flinn, M., and England, B. G. (1995). 'Childhood stress and family envi-ronment.' *Current Anthropology* 36: 854–66.

Grayson, D. K. (1994). 'Differential mortality and the Donner party dis-aster.' *Evolutionary Anthropology* 2: 151–9.

Legget, R., F., and Northwood, T. D. (1960). 'Noise surveys at cocktail parties.' *Journal of the Acoustical Society of America* 32: 16–18.

McCormick, N. B., and McCormick, J. W. (1992). 'Computer friends and foes: content of undergraduates' electronic mail.' *Computers and Human Behaviour* 8: 379–405.

Milardo, R. M. (ed.) (1988). *Families and Social Networks*. Sage, Newbury Park (Ca.).

Morgan, E. (1990). *The Scars of Evolution*. Souvenir Press, London.

Young, M., and Willmott, P. (1957). *Family and Kinship in East London*. Routledge and Kegan Paul, London.

Index

Abbott, David, 42
accents, 183–4, 186, 188
Ache hunter-gatherers, 186
'acoustic hiding', 92–3
advertising, sexual, 151, 177, 180–82,
    185, 187, 191, 192, 193
Afar desert, 108
'African Eve', 161
afterlife, 111
Age of the Dinosaurs, 11, 12
agriculture: discovery of, 170; see also
    farming
Aiello, Leslie, 111, 113, 124, 190
'alliance fickleness', 25
alliances: of apes, 19, 21, 151; coali-
    tion size, 66, 67; and grooming, 20,
    21, 22, 35, 44, 45, 46, 65, 67, 68;
    and harassment, 43–4; of monkeys,
    19, 21, 151; with neighbouring
    groups, 119; and primates, 19, 23,
    43–4, 66; and relatives, 166; and
    social inference, 22–3
altruism, 164, 165; reciprocal, 195
Amboseli National Park, Kenya, 21,
    101
American Indians, 117
Andersson, Malte, 189
animals: behaviour, 34; caged, 37; and
    familiarity, 168; and language, 51;
    as machines, 82, 83; and the mind,
    82; and mirror test for self-aware-
    ness, 91; tactical deception by, 92–5
antelope, 31, 108, 126, 176, 180
anthropology, 192
apes: ability to calculate the likely
    effect of their actions, 25; alliances,
    19, 21, 151; as anthropoid primates,

12, 63; brain size/body weight, 58;
    chest shape, 133, 134; common
    ancestor of monkeys and apes, 11;
    comparing understanding of anoth-
    er's intentions with that of mon-
    keys, 98–9; contact calls, 115; use
    of gestures, 135; and grooming, 4,
    21, 65, 77; and groups, 71, 112;
    and hairlessness, 107–8; heat stress
    and walking upright, 107; and
    intensionality, 91, 104; in marginal
    habitats, 14, 15, 106, 107, 109;
    monkeys out–compete, 13, 14, 15,
    106; neocortex size and group size,
    63; and observation of others'
    behaviour, 79; Old World, 11, 35,
    115, 140; and predation, 109; pub-
    lic behaviour, 7; social knowledge,
    60; social world, 1–3, 4, 9, 28; and
    theory of mind, 90–91; vocaliza-
    tions, 135, 140; and vowel sounds,
    141; see also chimpanzees; gibbons;
    gorillas; orang-utans
Aristotle, 30
arithmetical skills, 55, 90
Arnhem Zoo, 24, 151
art objects, 110
artificial keyboard language, 53, 93
ASL (deaf-and-dumb language), 52
Asperger's Syndrome, 88, 89, 90
Austen, Jane, 184, 185
Australian Aboriginals, 71, 117,
    153–4
australopithecines, 113, 127, 130
autism, 88–90, 95, 100

Babel, Tower of, 152–3, 154, 170

Index